Introduction
to Graphene

Introduction
to Graphene
Chemical and Biochemical
Applications

Challa Vijaya Kumar and
Ajith Pattammattel

University of Connecticut, Storrs, CT, United States

ELSEVIER

Elsevier
Radarweg 29, PO Box 211, 1000 AE Amsterdam, Netherlands
The Boulevard, Langford Lane, Kidlington, Oxford OX5 1GB, United Kingdom
50 Hampshire Street, 5th Floor, Cambridge, MA 02139, United States

Notices

Knowledge and best practice in this field are constantly changing. As new research and experience broaden our understanding, changes in research methods, professional practices, or medical treatment may become necessary.

Practitioners and researchers must always rely on their own experience and knowledge in evaluating and using any information, methods, compounds, or experiments described herein. In using such information or methods they should be mindful of their own safety and the safety of others, including parties for whom they have a professional responsibility.

To the fullest extent of the law, neither the Publisher nor the authors, contributors, or editors, assume any liability for any injury and/or damage to persons or property as a matter of products liability, negligence or otherwise, or from any use or operation of any methods, products, instructions, or ideas contained in the material herein.

British Library Cataloguing-in-Publication Data
A catalogue record for this book is available from the British Library

Library of Congress Cataloging-in-Publication Data
A catalog record for this book is available from the Library of Congress

ISBN: 978-0-12-813182-4

For Information on all Elsevier publications
visit our website at https://www.elsevier.com/books-and-journals

Working together
to grow libraries in
developing countries

www.elsevier.com • www.bookaid.org

Publisher: John Fedor
Acquisition Editor: John Fedor
Editorial Project Manager: Tasha Frank
Production Project Manager: Vijayaraj Purushothaman
Cover Designer: Christian J. Bilbow

Typeset by MPS Limited, Chennai, India

Contents

About the Authors xi

CHAPTER 1 Discovery of graphene and beyond...........................1
    1.1 Summary...1
    1.2 A Short History 2
    1.3 Wonder Material 4
    1.4 Terminology 4
    1.5 Comparisons With Other Carbon-Based Nanomaterials.............5
        1.5.1 Graphene Oxide 6
        1.5.2 Reduced Graphene Oxide 6
        1.5.3 Exfoliated Graphene 8
    1.6 Beyond Graphene..8
        1.6.1 Boron Nitride 8
        1.6.2 Transition Metal Dichalcogenides........................... 9
        1.6.3 MXenes 10
        1.6.4 Phosphorene 11
    1.7 Future Advances and Challenges.......................................11
        References... 12

**CHAPTER 2 Synthetic routes to graphene preparation
from the perspectives of possible biological
applications**... 17
    2.1 Summary...17
    2.2 Introduction 18
    2.3 Bottom-Up Approaches..19
        2.3.1 Chemical Vapor Deposition Method........................ 20
        2.3.2 Chemical Synthesis Method 21
    2.4 Top-Down Approaches 23
        2.4.1 Mechanical Peeling... 23
        2.4.2 Liquid-Phase Exfoliation of Graphene..................... 23
        2.4.3 Exfoliating Agents 27
    2.5 Conclusions 38
        References... 40

CHAPTER 3   Characterization techniques for graphene 45
3.1 Summary...45
3.2 Introduction 46
3.3 Spectroscopic Methods 46
    3.3.1 Absorption/Extinction Spectroscopy 47
    3.3.2 Raman Spectroscopy...50
    3.3.3 IR Spectroscopy 55
    3.3.4 X-Ray Photoelectron Spectroscopy 56
    3.3.5 X-Ray Diffraction (XRD)..57
    3.3.6 Zeta Potential Measurements..59
    3.3.7 Surface Plasmon Resonance Spectroscopy 61
3.4 Microscopy Methods..63
    3.4.1 SEM and TEM 63
    3.4.2 Atomic Force Microscopy/Scanning Tunneling
          Microscopy 65
    3.4.3 Fluorescence Quenching Microscopy............................ 67
3.5 Conclusions 67
    References... 69

CHAPTER 4   Inorganic analogues of graphene: synthesis, characterization, and applications 75
4.1 Summary...75
4.2 Introduction 76
4.3 Hexagonal Boron Nitride 77
    4.3.1 Production 78
    4.3.2 Characterization Techniques..79
    4.3.3 Applications of h-BN... 82
4.4 Transition Metal Dichalcogenides 83
    4.4.1 Exfoliation.. 83
    4.4.2 Characterization Techniques.. 85
    4.4.3 Applications 86
4.5 MXenes...87
    4.5.1 Production 88
    4.5.2 Characterization Techniques.. 90
    4.5.3 Applications 91
4.6 Monoatomic Layers..91
    4.6.1 Synthesis or Exfoliation.. 92
    4.6.2 Characterization 93
    4.6.3 Applications 94
4.7 Conclusions 94
    References... 95

CHAPTER 5   Graphene composites with inorganic 2-D materials.. **103**
  5.1 Summary...103
  5.2 Introduction 104
  5.3 Graphene Composites With h-BN 105
    5.3.1 Design and Preparations 105
    5.3.2 Characterization 108
    5.3.3 Properties and Applications 110
  5.4 Graphene Composites With MoS_2......................110
    5.4.1 Design and Preparation................................110
    5.4.2 Characterization 112
    5.4.3 Properties and Applications 113
  5.5 Composites With Other Inorganic Layered Materials 114
    5.5.1 Composites With TMDs 114
    5.5.2 Nongraphene Heterostructures.......................117
  5.6 Conclusions 117
    References...118

CHAPTER 6   Graphene hybrids with carbon allotropes............... **123**
  6.1 Summary...123
  6.2 Introduction 124
  6.3 Graphene/CNT Hybrids 125
    6.3.1 Covalent CNT/Graphene Hybrids 125
    6.3.2 Noncovalent CNT/Graphene Hybrids 127
  6.4 Graphene/Fullerene Hybrids 129
  6.5 Organic Chemistry of Graphene 129
  6.6 Conclusions 134
    References...135

CHAPTER 7   Graphene composites with synthetic polymers....... **141**
  7.1 Summary...141
  7.2 Introduction 142
  7.3 In situ Synthesis of Graphene/Polymer Composites 144
    7.3.1 Design and Preparation................................144
    7.3.2 Characterization 144
    7.3.3 Properties and Applications 146
  7.4 Post-modification Approaches 147
    7.4.1 Design and Preparation................................147
    7.4.2 Characterization 147
    7.4.3 Properties and Applications 149

7.5 Conclusions 150
References... 150

**CHAPTER 8 Graphene composites with proteins
and biologics** **155**
  8.1 Summary..155
  8.2 Introduction 156
  8.3 Design and Preparation of Graphene Biohybrids.....................159
    8.3.1 Covalent and Noncovalent Approaches 159
  8.4 Characterization of Composites....................................162
    8.4.1 Equilibrium Binding Studies 163
    8.4.2 Fluorescence Spectroscopy 165
    8.4.3 Zeta Potential Studies 166
    8.4.4 Atomic Force Microscopy 167
    8.4.5 X-Ray Diffraction 168
    8.4.6 CD Spectroscopy...168
    8.4.7 Biological Activity Studies...................................171
  8.5 Biocatalysis Using Graphene Platform...............................171
    8.5.1 Preservation of Structure and Function of Bound
Enzymes 172
    8.5.2 Thermodynamic Stabilities of Enzymes.........................174
    8.5.3 Kinetic Stability of Enzymes 175
  8.6 Biosensing 177
    8.6.1 Fluorescence-Based Graphene Biosensors 177
    8.6.2 Electrochemical Graphene Sensors 178
  8.7 Drug Delivery Applications 179
  8.8 Conclusions 179
References... 180

CHAPTER 9 Nanotoxicity of graphene **187**
  9.1 Summary..187
  9.2 Introduction 188
  9.3 Challenges in Experimental Design of Nanotoxicity Assays ...189
  9.4 Methods for Nanotoxicity Assessments................................192
  9.5 In Vitro Toxicity of Graphene 192
  9.6 In Vivo Toxicity of Graphene 198
  9.7 Conclusions 198
References... 202

CHAPTER 10 Future of graphene revolution and roadmap 207

    10.1 Summary..207

    10.2 Introduction 208

    10.3 Graphene Manufacturing: A Critical Assessment 209

    10.4 Health and Environmental Risk Assessment of Graphene 210

    10.5 Graphene: Fundamental Education and Society........................211

    10.6 Conclusions 211

         References... 212

Index 215

About the Authors

Ajith Pattammattel was born in Cochin, India. He earned his bachelor's degree in Chemistry from Mahatma Gandhi University and master's degree in Applied Chemistry from Cochin University of Science and Technology. In 2011, he joined Dr. Challa V. Kumar's research lab at the University of Connecticut for doctoral studies. During this period, he published several papers relating to graphene in biological media. After receiving a Ph.D. in Biological Chemistry in 2016, he moved to University of California-Merced for postdoctoral studies. His postdoctoral studies focus on speciation of nanoparticulates in the environment using advanced electron microscopy and synchrotron X-ray spectroscopy. Dr. Pattammattel's research interests include nanotechnology, biophysical chemistry, nanotoxicity, and environmental chemistry. During his free time, he likes to read, hike, and cook.

Prof. Challa Vijaya Kumar was born in a small village, Tenali, in southern India. After graduation from the New Science College, he received his MSc degree from the Osmania University, Hyderabad, and went on to earn a Ph.D. in Chemistry from the Indian Institute of Technology, Kanpur. Then, he moved to the U.S. and continued his research as a postdoctoral fellow in photochemistry at the Radiation Laboratory, University of Notre Dame, and later as Research Associate at Columbia University where he made an important discovery of DNA charge transport in the laboratories of Professors Barton and Turro. He has been faculty at the University of Connecticut since 1988 in the Chemistry Department and currently holds joint appointments with the Department of Molecular and Cell Biology and the Institute of Materials Science.

He received eight research grants from the National Science Foundation, one from the Department of Energy, and another from NASA, as well as numerous research grants from foundations and industrial collaborators. One of his recent discoveries led to the commercialization of protein-based nanoparticles (GlowDots). He has received several awards including a Merck Faculty Development Award, Fulbright Research Excellence Award, and AAUP Research Excellence Award. He has published 170 peer-reviewed research papers in major journals such as *Science, Proceedings of National Academies of Sciences U.S., Journal of the American Chemical Society, Angewandte Chemie*, as well as several Elsevier and RSC journals. His research work has been cited more than 8000 times with an h-index of approximately 38.

Drawn to artistic endeavors at an early age, he played in dramatic arts celebrations in his hometown. An amateur watercolorist, his office in the Chemistry Department showcases several of his watercolors. He enjoys listening to Indian bamboo flute and mandolin, primarily South Indian classical music, and loves to garden. He is highly passionate about "Art-in-Science" and has "Art-in-Nanochemistry" exhibit displayed at many galleries throughout Connecticut.

Discovery of graphene and beyond

1

CHAPTER OUTLINE

1.1 Summary 1
1.2 A Short History 2
1.3 Wonder Material 4
1.4 Terminology 4
1.5 Comparisons With Other Carbon-Based Nanomaterials 5
    1.5.1 Graphene Oxide 6
    1.5.2 Reduced Graphene Oxide 6
    1.5.3 Exfoliated Graphene 8
1.6 Beyond Graphene 8
    1.6.1 Boron Nitride 8
    1.6.2 Transition Metal Dichalcogenides 9
    1.6.3 MXenes 10
    1.6.4 Phosphorene 11
1.7 Future Advances and Challenges 11
References 12

1.1 SUMMARY

The advent of graphene as a wonder material of the 21st century is an inspiring story. This chapter focuses on a brief history of graphene before and after the isolation and complete characterization by Geim and Novoselov in 2004. Interestingly, graphite chemistry began in 1859, when the syntheses of graphene oxide (GO) and reduced graphene oxide (rGO) were reported. Microscopic images of single-layer graphene (SLG) oxide sheets were observed in 1948, but the amazing properties of graphene got attention only in the 21st century. After the successful isolation of graphene, other atomically thin layers of inorganic materials were reported, such as hexagonal boron nitride (h-BN), molybdenum sulfide, and other dichalcogenides. A short introduction to these graphene family members is given in Section 1.6. Finally, the challenges and opportunities of these materials, such as scalability, cost, and toxicity, are examined. This chapter gives an introduction to the two-dimensional materials, which are going to be the foci of the following chapters.

Introduction to Graphene. DOI: http://dx.doi.org/10.1016/B978-0-12-813182-4.00001-5
© 2017 Elsevier Inc. All rights reserved.

1.2 A SHORT HISTORY

Graphene is the latest stable carbon allotrope where the sp^2 carbon atoms are oriented in a two-dimensional honeycomb lattice to form atomically thin layers. The bonding between the adjacent carbons and the delocalization of the bonding electrons throughout the two-dimensional (2-D) lattice contribute to most of the interesting properties of graphene. The term "graphene" originated when studying the intercalation compounds of the parent material graphite,[1] which is the three-dimensional, naturally occurring form of graphene (Fig. 1.1A). Graphite, which means *"to write"* (in Latin) has been known for 500 years and is widely used in pencils (in combination with clays). It is made of graphene layers where the layers are stacked with an interlayer distance of 3.34 Å.[2] Graphene is considered as the ultimate polycyclic aromatic hydrocarbon (including naphthalene, anthracene, pyrene, etc.), and thus, the name "graph-ene" has been suggested (Fig. 1.1B and C).[3–5]

According to early theoretical calculations, 2-D atomic crystals such as graphene were considered to be thermodynamically unstable.[6,7] Studies predicted that the 2-D atomic crystals are stable only in a 3-D from as part of a crystal structure, just like they have been seen to be in graphite or thin films grown on

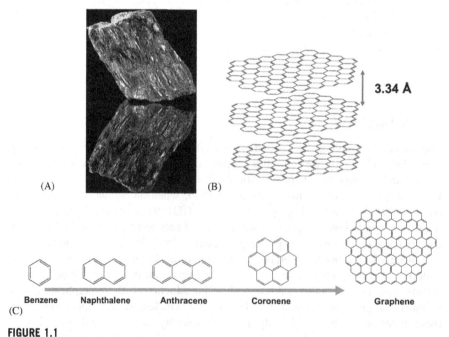

(A) (B)

(C) Benzene Naphthalene Anthracene Coronene Graphene

FIGURE 1.1

(A) Graphite crystal (©Anneka/Shutterstock). (B) 3-D packing of graphene sheets in graphite. (C) Hierarchy of aromatic hydrocarbons from benzene to graphene.

solid supports.[8] Calculations showed that the melting point of atomic layers decreases with a decrease in the number of layers, so eventually they segregate into islands of multilayered structures or they decompose at room temperature.[9] This common wisdom was discredited in 2004 by Geim and Novoselov, physicists at the University of Manchester, after their successful isolation of atomically thin, stable free-standing carbon nanosheets from graphite using the "Scotch Tape" method. They were honored with the 2010 Nobel Prize for Physics![10] This discovery has led to an explosive growth of graphene research ever since.

Although graphene became popular in the last decade after the invention by Geim and Novoslov, the "hunt" for single-layer carbon sheets began back in the 19th century. Sir Benjamin Collins Brodie, at the University of Oxford, in 1859 reported the lamellar structure of graphite (in a paper titled *On Atomic Weight of Graphite*).[11] According to Brodie's original work, upon oxidation, graphite forms "carbonic acid," which is a light-yellow-colored substance (this substance is now known as GO). GO is often referred to as graphene in the current literature, incorrectly, and the distinction is not very clear. Thermal reduction of "carbonic acid" to form rGO (another popular material currently being investigated extensively) was also noted in Brodie's work for the first time.[11] Transmission electron microscopy images of a few nanometers thick GO sheets were published in 1948 by Ruess and Vogut (Fig. 1.2A), but these sheets were crumpled and folded forms of graphene multilayers.[12] Monolayer GO sheets were first imaged in 1962 by Boehm et al.[13] and a more recent scanning probe microscopy image is shown in Fig. 1.2B, where the positions of the carbon atoms are readily visualized to be 1.42 Å apart.[14]

SLG grown on metal surfaces, known as epitaxial graphene, gained much attention in the 1970s, but the electronic interaction of the graphene π-electrons with those of the underlying metal support significantly affected the graphene properties.[16] Attempts were made to separate graphene sheets from graphite crystals, but isolation of monolayers was unsuccessful, until the 2004 discovery of the Scotch Tape method.[4]

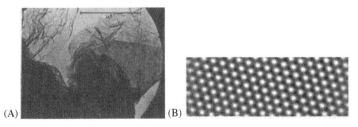

(A)   (B)

FIGURE 1.2

(A) Transmission electron microscopy image of graphene oxide published in 1948 by Ruess and Vogut.[12] (B) Scanning tunneling microscopy image of graphene on gold surface (15×5 nm^2 frame).[15]

(A) Reproduced with permission from Springer. (B) Reproduced with permission from Elsevier.

1.3 WONDER MATERIAL

Since 2004, graphene has attained unprecedented importance in both academic and industrial research as well as in applications. The global market for graphene use in electronics, semiconductors, and battery applications is growing rapidly, as more efficient, large-scale production methods are continuously being discovered. Graphene has very interesting electronic and material properties when compared to any existing material.[10] SLG, which is one atom thick, is optically transparent, electrically conductive, and flexible, which makes it an ideal platform for the manufacturing of displays and flexible electronic devices. The theoretical limit of electron mobility in graphene due to the conjugated π-electrons is 200,000 cm^2/V s (at $\sim$5K), which is very high when compared to that of Si with a much smaller value of 1500 cm^2/V s.[17] The electrons in graphene are described as massless particles traveling without scattering, which is called ballistic transport, and this has been suggested to be responsible for its high carrier mobility.[18] However, structural defects, edges, and substrate effects limit the magnitude of observed carrier mobility due to electron scattering.[19]

Graphene is the strongest material known to man, with a tensile strength of $\sim$130 GPa, which is much greater than that of steel (0.4 GPa).[20] The attractive nature of graphene strength is combined with its light weight and high theoretical surface area of 2000 m^2/g. Graphene also shows elastic properties with a large Young's modulus of 500 GPa, while that of rubber is only 0.01−0.1 GPa.[20] The high strength, high electrical conductivity, light weight, and elastic nature of graphene are unique properties which are attractive for its use in flexible electronics.[21] However, these attributes are for graphene with no defects or for samples of very low-defect levels, and these high values are often not realized.[22]

Promises of a "graphene revolution" vary from ultrafast electronic devices to water purification systems, giving it both scientific and economic relevance.[23] The developments in these numerous applications depend on the production, cost, characterization, fabrication, and safety of the final material. The following sections will provide an overview with foci on these aspects of graphene. The major emphasis in this book will be on the solution-phase chemistry and biological applications of graphene. The nontoxicity of graphene and its derivatives will also be covered in this book.

1.4 TERMINOLOGY

The term "graphene," often used to represent graphene derivatives, rather than graphene, leads to massive confusion in the literature.[24] GO and rGO (Fig. 1.3), which possess different chemical and mechanical properties when compared to graphene, are often called "graphene" in the literature.[24] The term "pristine graphene" is used for graphene that was not produced by oxidation or reduction

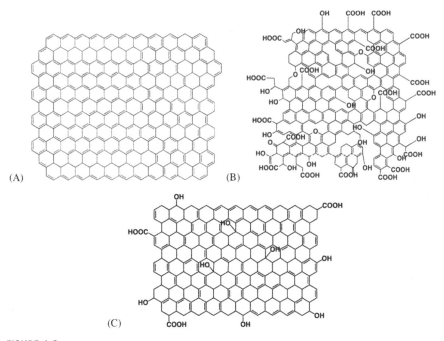

FIGURE 1.3

Representative structures of (A) graphene, (B) graphene oxide, and (C) reduced graphene oxide.

reactions. In this book, "graphene (Gr)" will be exclusively used for graphene derived directly from graphite, while GO or rGO will be specified as such.

1.5 COMPARISONS WITH OTHER CARBON-BASED NANOMATERIALS

Carbon nanotubes (CNTs) and fullerenes are allotropic carbon forms on a nanoscale, discovered in the early 1990s,[25] whereas bulk phase diamond has been known for centuries. Both CNTs and fullerenes (Fig. 1.4) proved to be useful for electronic applications, but their limited availability and high cost outweigh their performance. Graphene can be prepared from graphite, on the other hand, which is abundant and inexpensive, and it is an ideal candidate for numerous applications. However, the isolation of defect-free graphene from graphite on an industrial scale is yet to be achieved. However, the synthesis of GO,[26] which is the oxidized form of graphene is possible on a large-scale. Graphene oxide and rGO are useful in the preparation of composite materials and for electronic applications.[27] Large-scale preparation of graphene directly from graphite is also

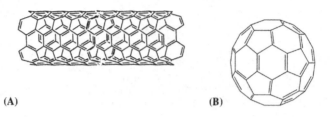

(A) **(B)**

FIGURE 1.4

Schematic structures of (A) carbon nanotube and (B) fullerene (C_{60}).

possible, as suggested by the recent literature, and this aspect is the major thrust of this introductory book.

1.5.1 GRAPHENE OXIDE

As the name indicates, GO is the oxidized derivative of graphene, which is obtained by the oxidation of graphite using strong oxidants. The safe, well-established method for producing GO was reported by Hummers in 1958.[26] Hummers method was cited about 100 times until the report of graphene isolation from graphite in 2004, which started a revolution in graphene research and, in 2016, Hummers method had 15,000 citations (Google Scholar database). Hummers method uses a mixture of permanganate and sulfuric acid (Fig. 1.5) to partially oxidize the material, and it replaced the use of perchlorate in the Staudenmeir–Hoffman–Hamdi oxidation method,[28] for improved safety.[29]

The formation of GO involves intercalation of sulfuric acid into graphite, by exposure of graphite to an acidic oxidizing medium, followed by oxidation of carbon layers.[31] The final step is the hydrolysis of the intermediate to produce different functional groups, depending on the position of the carbon atom that has been oxidized.[30] Graphene oxide thus prepared has numerous defects, as a consequence of harsh oxidation, but the defect sites are decreased by the chemical/electrochemical reduction of GO.[32] More recently, nitrate in the Hummers method was replaced by persulfate and this avoids the release of nitrogen oxides and results in completely exfoliated GO.[33]

1.5.2 REDUCED GRAPHENE OXIDE

Highly oxidized GO is reduced by chemical or physical treatments and results in partially deoxygenated sheets, called rGO (Fig. 1.6).[29] A commonly used reducing agent for GO preparation is hydrazine hydrate (NH_2-NH_2), which is highly toxic.[34] Alternatively, mild reducing agents such as sodium borohydride ($NaBH_4$), vitamin C, reducing sugars, urea, photochemical methods, or linear voltammetry were used to produce rGO from GO.[32] The electronic properties of rGO, such as conductivity and supercapacitance, are improved when compared

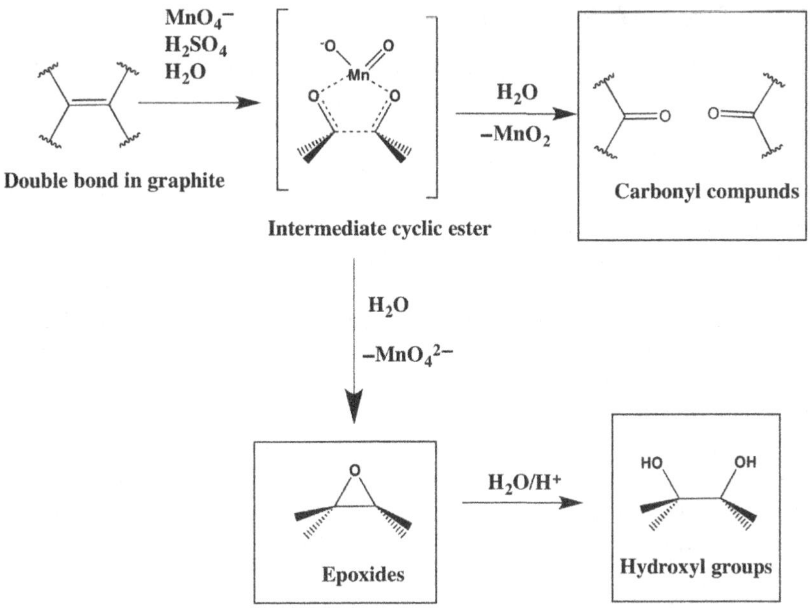

FIGURE 1.5

Formation of graphene oxide from graphite by strong oxidation by the Hummers method.[30]

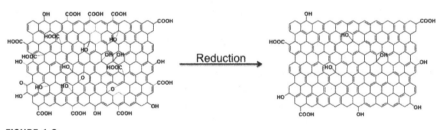

FIGURE 1.6

Reduction of graphene oxide to reduced graphene oxide by physical, chemical, photochemical, or electrochemical methods.

to that of GO, but lower dispersivity of rGO limits its solution processability.[35] Graphene oxide was reduced on a compact disk using LASER scribbling which showed high supercapacitance and led to another branch of the research area.[36] However, the presence of defects and multistep chemical processing to yield rGO is not convenient for large-scale applications. Nearly defect-free graphene can be isolated from graphite without oxidation/reduction by other means described below.[37]

1.5.3 EXFOLIATED GRAPHENE

Graphene is mass produced by micromechanical forces leading to the delamination and fragmentation of multilayer graphite, under favorable conditions.[37] The first demonstration of exfoliation of graphite to form graphene suspended in a solvent was reported in 2008, using *N*-methyl pyrrolidine (NMP) as the medium, aided by ultrasonication.[38] To date, numerous exfoliation methods and exfoliation agents are known to produce graphene from graphite (Chapter 2: Synthetic routes to graphene preparation from the perspectives of possible biological applications is dedicated to discussing this topic in detail). Compared to GO or rGO, exfoliated graphene has many fewer defects, which is highly desirable for electronic applications. Moreover, minimum chemical processing is favorable for large-scale production without toxic byproducts. Organic solvents such as NMP are often used, which are toxic, but exfoliation in water (biological solvent) using several biocompatible agents including proteins, DNA, and carbohydrates promises green and inexpensive production of high-quality graphene.[39,40] These methods are also applied to exfoliate several other two-dimensional materials, other than graphene, as discussed below.[41]

1.6 BEYOND GRAPHENE

After the discovery of graphene in 2004, a distinct branch of research in 2-D materials rapidly evolved. Atomically thin layers of several such materials, with high surface area and specific chemical compositions, showed superior properties when compared to the corresponding bulk materials. These 2-D materials are being extensively tested for applications in electronics, catalysis, and biomedicine.[41] Detailed descriptions of these materials and their applications will be given in Chapter 4, Inorganic analogues of graphene: synthesis, characterization and applications and Chapter 5, Graphene composites with inorganic 2-D materials, but a short introduction is given here. The "hunt" for other, stable, thin atomic layers, other than graphene has began with *h*-BN nanolayers (discussed below).[42]

1.6.1 BORON NITRIDE

Boron nitride (BN), consisting of boron−nitrogen covalent bonds, was commonly used as a refractory material. Isoelectronic to sp^2 carbon lattice, BN was generally compared with carbon allotropes. The cubic form of BN (c-BN) has a diamond-like crystalline arrangement and the bulk crystal of *h*-BN is analogous to graphite crystal.[43] The 2-D sheets of *h*-BN are the most stable and soft among its polymorphs, and bonding in *h*-BN is similar to that in aromatic compounds, but its considerably less covalency and higher ionic character make it one of the best proton conductors but also an electrical insulator. Its thermal conductivity is the

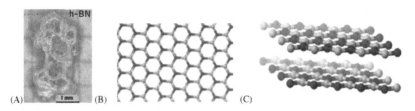

FIGURE 1.7

Hexagonal boron nitride (*h*-BN). (A) Picture of bulk *h*-BN; (B) and (C) Structure of *h*-BN layers.

©*Wikipedia.org. Public domain Pictures.*

highest among all electrical insulators (Fig. 1.7). Atomically thin *h*-BN sheets, also called "white graphene" can be synthesized by chemical vapor deposition (CVD) of molecular precursors, such as ammonia−borate.[44] Exfoliation of bulk *h*-BN under suitable conditions was also demonstrated for large-scale applications in coatings and cosmetics including, but not limited to, lipsticks and lip balms. *h*-BN is used as a substrate to grow large-area graphene films because of its low lattice mismatch with graphene (1.7%). Nanolayers of *h*-BN display excellent thermal stability, chemical inertness, and high optical transparency, when compared with those of graphene. In contrast to electronically conductive graphene, *h*-BN layers are insulators (band gap ∼6 eV) because of the absence of the π-electrons and they show fire-retardant abilities.[45] The layers of *h*-BN have unusually high proton conduction rates and when combined with high electrical resistance, these could be useful for fuel cell applications.[46] Hence, inorganic analogues of graphene, such as *h*-BN, have paved the way to discover atomic layers of other elements with tunable properties and these include transition metal dichalogenides (TMDs) which are described next.[47]

1.6.2 TRANSITION METAL DICHALCOGENIDES

As the name indicates, TMDs are formed from a transition metal and group 16 elements (oxygen group: chalcogens). These compounds with general formula MX_2 (M, metal; X, halogen) have zig-zag bonding pattern with X−M−X bonds, where the chalcogen has out-of-plane orientation with respect to the metal plane (Fig. 1.8).[48] Oxides, sulfides, selenides, and tellurides of transition metals such as molybdenum (Mo) and tungsten (W) form layered structures similar to graphite.[48] A wide variety of TMDs are known to exist in monolayer form and the most widely studied member is molybdenum sulfide (MoS_2) because of its availability in bulk form and a band gap that is comparable to many semiconductors (band gap of 1.2−1.9 eV), depending on the number of layers.[49] TMD monolayers can be synthesized by CVD, exfoliation from bulk and molecular beam epitaxy techniques. Apart from the chalcogens, nitrides, carbides, or carbonitrides of transition

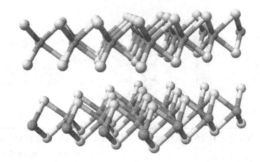

FIGURE 1.8

MoS_2 monolayer structure

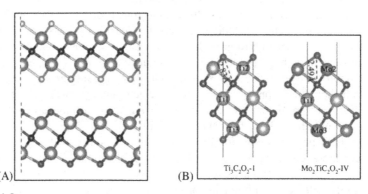

FIGURE 1.9

Bonding in Ti_2CF_2 (top) and Ti_2CO_2. Most energetically favorable structures of $Ti_3C_2O_2$ (A) and $Mo_2TiC_2O_2$ (B).

metals also form stable monolayer structures, known as MXenes (described briefly next).[50]

1.6.3 MXENES

MXenes are 2-D layers derived from MAX phases, where M is a transition metal (Ti, V, Cr, etc.), A is a group A element (e.g., Si, Al, etc., Groups 1, 2, 12, 13, 14, 15, 16, 17 in current IUPAC naming) and X represents carbon and/or nitrogen.[51] They have the general formula $M_{n+1}AX_n$, where $n = 1$, 2, or 3, for example, Ti_3AlC_2. Nanolayers of MAX phases are achieved by acid etching (usually hydrofluoric acid) of the bulk material to yield composition of $M_{n+1}X_n$ (Fig. 1.9), named as a graphene family member, MXenes (e.g., Ti_3C_2).[50] MXenes find versatile applications in supercapacitors, electromagnetic interference shielding, batteries, and many more.[51]

1.6.4 **PHOSPHORENE**

A single-atom-layer allotrope of phosphorous, analogous to graphene was also realized in 2014, by micromechanical exfoliation of black phosphorous.[52] Unlike graphene, phosphorene has a band gap of $\sim 2.0\,\text{eV}$, perfect for semiconducting applications when compared to that of silicon ($\sim 3\,\text{eV}$).[53] Liquid-phase exfoliation methods are also established for phosphorene, which promises its large-scale applications.[54] Since phosphorene is highly reactive to oxygen, deoxygenation of solvents and stabilizing agents or suitable solvent is needed.[55] Phosphorene is stable in water in the absence of oxygen. Regardless of its sensitivity to air, phosphorene was demonstrated to be useful in transistors, sensors, and catalysis.[55] Phosphorene, like single-atom layers of boron (borophene),[56] antimony (antomonene),[57] and so on, are known to exist, but stability and scalable production are still to be realized for practical applications.

1.7 **FUTURE ADVANCES AND CHALLENGES**

The advent of graphene catalyzed the formation of an interesting class of materials: the −ene materials family. They range from single-atom thin layers to hetero atom derivatives, with ample opportunities of band gap tuning and chemical functionalization.[58] One of the key challenges in this field is the production of high-purity materials, at affordable prices. Liquid-phase exfoliation offers low-cost, high-yield production, compared to CVD methods, but fragmentation of the sheets affects their mechanical, chemical, and electronic properties.[59] Also, the bulk materials are not naturally occurring for many two-dimensional materials, and these require additional synthesis methods to be developed.[60] The challenge to produce high-quality and large-area sheets, inexpensively, needs to be addressed in the future. Most of these materials primarily target the preparation of transparent, flexible, high-speed electronic components and energy-storage materials.[60] The poor stability of some 2-D materials in air or water restricts their uses in research laboratories, which is a roadblock for real-life applications.[61] Approaches to stabilize those materials by coating or by other means could boost this field. However, several applications of graphene and other 2-D materials in catalysis and biology have already been demonstrated and many laboratories are still exploring such applications.[62]

Biological applications of 2-D materials depend on the stability of these materials in biological media and their respective toxicity levels to biological systems. As mentioned above, some of the 2-D materials are reactive to water and are therefore not suitable for biological applications because of the ubiquitous presence of water in all known biological systems.[63] Graphene, GO, rGO, h-BN, MoS_2, and MXenes have been used in biology for drug delivery, antibacterial properties, photodynamic therapy, and so on. The stability and toxicity of these materials to cells or organisms are only beginning to be evaluated and these are important to harness the unique features of these materials. Nevertheless, graphene family materials put forward an unprecedented amount of theoretical and application-oriented research in the fields of physics, chemistry, material sciences, and biology, but they are yet to realize applications in everyday life.

REFERENCES

1. Mouras, S., et al. Synthesis of First Stage Graphite Intercalation Compounds with Fluorides. *Rev. Chim. Minér.* **1987,** *24,* 572.
2. Bernal, J. D. The Structure of Graphite. *Proc. R. Soc. Lond* **1924,** *A106* (740), 749−773.
3. Boehm, H. P.; Setton, R.; Stumpp, E. Nomenclature and Terminology of Graphite Intercalation Compounds. *Pure Appl. Chem.* **1994,** *66* (9), 1893−1901.
4. Novoselov, K. S.; Geim, A. K.; Morozov, S. V.; Jiang, D.; Zhang, Y.; Dubonos, S. V.; Grigorieva, I. V.; Firsov, A. A. , Electric Field Effect in Atomically Thin Carbon Films. *Science* **2004,** *306* (5696), 7088−7098.
5. Boehm, H. P.; Clauss, A.; Fischer, G. O.; Hofmann, U. Das Adsorptionsverhalten sehr dünner Kohlenstoff-Folien. *Z. Anorg. Allg. Chem.* **1962,** *316* (3−4), 119−127.
6. Geim, A. K.; Novoselov, K. S. The Rise of Graphene. *Nat. Mater.* **2007,** *6* (3), 183−191.
7. This Month in Physics History: October 22, 2004: Discovery of Graphene. APS News Ser. II 2009, 18 (9), 2.
8. Forbeaux, I.; Themlin, J. M.; Debever, J. M. Heteroepitaxial Graphite on 6H-SiC (0001) Interface Formation Through Conduction-Band Electronic Structure. *Phys. Rev. B* **1998,** *58* (24), 16396−16406.
9. Dreyer, D. R.; Ruoff, R. S.; Bielawski, C. W. From Conception to Realization: An Historical Account of Graphene and Some Perspectives for Its Future. *Angew. Chem. Int. Ed.* **2010,** *49* (49), 9336−9344.
10. Geim, A. K. Graphene Prehistory. *Phys. Scripta* **2012,** *2012* (T146), 014003.
11. Brodie, B. C. On the Atomic Weight of Graphite. *Philos. Trans. R. Soc. Lond.* **1859,** *149,* 249−259.
12. Ruess, G.; Vogt, F. Höchstlamellarer Kohlenstoff aus Graphitoxyhydroxyd. *Monatshefte für Chemie und verwandte Teile anderer Wissenschaften* **1948,** *78* (3), 222−242.
13. Boehm, H. P.; Clauss, A.; Fischer, G. O.; Hofmann, U. Das Adsorptionsverhalten sehr dunner Kohlenstoff-Folien. *Anorg. Allg. Chem.* **1962,** *316,* 119−127.
14. Sakai K-i, T. K.; Fukui K-i.; Nakanishi, T.; Enoki, T. Honeycomb Superperiodic Pattern and Its Fine Structure Near the Armchair Edge of Graphene Observed by Low-Temperature Scanning Tunneling Microscopy. *Phys. Rev. B* **2010,** *81* (23), 235417.

15. Schouteden, K.; Galvanetto, N.; Wang, C. D.; Li, Z.; Van Haesendonck, C. Scanning Probe Microscopy Study of Chemical Vapor Deposition Grown Graphene Transferred to Au(111). *Carbon* **2015,** *95,* 318−322.
16. Oshima, C.; Nagashima, A. Ultra-thin Epitaxial Films of Graphite and Hexagonal Boron Nitride on Solid Surfaces. *J. Phys. Condens. Matter* **1997,** *9,* 1−20.
17. Bolotin, K. I.; Sikes, K. J.; Jiang, Z.; Klima, M.; Fudenberg, G.; Hone, J.; Kim, P.; Stormer, H. L. Ultrahigh Electron Mobility in Suspended Graphene. *Solid State Commun.* **2008,** *146* (9−10), 351−355.
18. Novoselov, K. S.; Geim, A. K.; Morozov, S. V.; Jiang, D.; Katsnelson, M. I.; Grigorieva, I. V.; Dubonos, S. V.; Firsov, A. A. Two-Dimensional Gas of Massless Dirac Fermions in Graphene. *Nature* **2005,** *438* (7065), 197−200.
19. Chen, J.-H.; Jang, C.; Xiao, S.; Ishigami, M.; Fuhrer, M. S. Intrinsic and Extrinsic Performance Limits of Graphene Devices on SiO2. *Nat. Nano.* **2008,** *3* (4), 206−209.
20. Lee, C.; Wei, X.; Kysar, J. W.; Hone, J. Measurement of the Elastic Properties and Intrinsic Strength of Monolayer Graphene. *Science* **2008,** *321* (5887), 385−388.
21. Briggs, B. D.; Nagabhirava, B.; Rao, G.; Geer, R.; Gao, H.; Xu, Y.; Yu, B. Electromechanical Robustness of Monolayer Graphene with Extreme Bending. *Appl. Phys. Lett.* **2010,** *97* (22), 223102.
22. Zandiatashbar, A.; Lee, G.-H.; An, S. J.; Lee, S.; Mathew, N.; Terrones, M.; Hayashi, T.; Picu, C. R.; Hone, J.; Koratkar, N. Effect of Defects on the Intrinsic Strength and Stiffness of Graphene. *Nat. Commun.* **2014,** *5,* 3186.
23. Zurutuza, A.; Marinelli, C. Challenges and Opportunities in Graphene Commercialization. *Nat. Nano.* **2014,** *9* (10), 730−734.
24. Bianco, A.; Cheng, H.-M.; Enoki, T.; Gogotsi, Y.; Hurt, R. H.; Koratkar, N.; Kyotani, T.; Monthioux, M.; Park, C. R.; Tascon, J. M. D., et al. All in the Graphene Family − A Recommended Nomenclature for Two-Dimensional Carbon Materials. *Carbon* **2013,** *65,* 1−6.
25. Pujado, M. P. *Carbon Nanotubesas Platforms for Biosensors with Electrochemical and Electronic Transductions;* Springer: Heidelberg, 2012.
26. Hummers, W. S.; Offeman, R. E. Preparation of Graphitic Oxide. *J. Am. Chem. Soc.* **1958,** *80* (6), 1339.
27. Ray, S. C.; Jana, N. R. *Applications of Graphene and Graphene Oxide based Nanomaterials;* William Andrew Publishing: Oxford, 2015.
28. Ojha, K.; Anjaneyulu, O.; Ganguli, A. Graphene-Based Hybrid Materials: Synthetic Approaches and Properties. *Curr. Sci.* **2014,** *107* (3), 397−418.
29. Chua, C. K.; Pumera, M. Chemical Reduction of Graphene Oxide: A Synthetic Chemistry Viewpoint. *Chem. Soc. Rev* **2014,** *43* (1), 291−312.
30. Kang, J. H.; Kim, T.; Choi, J.; Park, J.; Kim, Y. S.; Chang, M. S.; Jung, H.; Park, K. T.; Yang, S. J.; Park, C. R. Hidden Second Oxidation Step of Hummers Method. *Chem. Mater.* **2016,** *28* (3), 756−764.
31. Dimiev, A. M.; Tour, J. M. Mechanism of Graphene Oxide Formation. *ACS Nano.* **2014,** *8* (3), 3060−3068.
32. Pei, S.; Cheng, H.-M. The Reduction of Graphene Oxide. *Carbon* **2012,** *50* (9), 3210−3228.
33. Chen, J.; Yao, B.; Li, C.; Shi, G. An Improved Hummers Method for Eco-Friendly Synthesis of Graphene Oxide. *Carbon* **2013,** *64,* 225−229.

34. Vernot, E. H.; MacEwen, J. D.; Bruner, R. H.; Haun, C. C.; Kinkead, E. R.; Prentice, D. E.; Hall, A.; Schmidt, R. E.; Eason, R. L.; Hubbard, G. B., et al. Long-Term Inhalation Toxicity of Hydrazine. *Fundam. Appl. Toxicol.* **1985,** *5* (6), 1050–1064.

35. Konios, D.; Stylianakis, M. M.; Stratakis, E.; Kymakis, E. Dispersion Behaviour of Graphene Oxide and Reduced Graphene Oxide. *J. Coll. Inter. Sci.* **2014,** *430,* 108–112.

36. El-Kady, M. F.; Strong, V.; Dubin, S.; Kaner, R. B. Laser Scribing of High-Performance and Flexible Graphene-Based Electrochemical Capacitors. *Science* **2012,** *335* (6074), 1326–1330.

37. Yi, M.; Shen, Z. A Review on Mechanical Exfoliation for the Scalable Production of Graphene. *J. Mater. Chem. A* **2015,** *3* (22), 11700–11715.

38. Hernandez, Y.; Nicolosi, V.; Lotya, M.; Blighe, F. M.; Sun, Z.; De, S.; McGovern, I. T.; Holland, B.; Byrne, M.; Gun'Ko, Y. K., et al. High-Yield Production of Graphene by Liquid-Phase Exfoliation of Graphite. *Nat. Nano.* **2008,** *3* (9), 563–568.

39. Pattammattel, A.; Kumar, C. V. Kitchen Chemistry 101: Multigram Production of High Quality Biographene in a Blender with Edible Proteins. *Adv. Funct. Mater.* **2015,** *25,* 7088–7098.

40. Paredes, J. I.; Villar-Rodil, S. Biomolecule-Assisted Exfoliation and Dispersion of Graphene and Other Two-Dimensional Materials: A Review of Recent Progress and Applications. *Nanoscale* **2016,** *8* (34), 15389–15413.

41. Butler, S. Z.; Hollen, S. M.; Cao, L.; Cui, Y.; Gupta, J. A.; Gutiérrez, H. R., et al. Progress, Challenges, and Opportunities in Two-Dimensional Materials Beyond Graphene. *ACS Nano.* **2013,** *7* (4), 2898–2926.

42. Corso, M.; Auwärter, W.; Muntwiler, M.; Tamai, A.; Greber, T.; Osterwalder, J. Boron Nitride Nanomesh. *Science* **2004,** *303* (5655), 217–220.

43. Wentorf, R. H., Jr Cubic Form of Boron Nitride. *J. Chem. Phys.* **1957,** *26* (4), 956.

44. Stehle, Y.; Meyer, H. M.; Unocic, R. R.; Kidder, M.; Polizos, G.; Datskos, P. G.; Jackson, R.; Smirnov, S. N.; Vlassiouk, I. V. Synthesis of Hexagonal Boron Nitride Monolayer: Control of Nucleation and Crystal Morphology. *Chem. Mater.* **2015,** *27* (23), 8041–8047.

45. Watanabe, K.; Taniguchi, T.; Kanda, H. Direct-Bandgap Properties and Evidence for Ultraviolet Lasing of Hexagonal Boron Nitride Single Crystal. *Nat. Mater.* **2004,** *3* (6), 404–409.

46. Hu, S., et al. Proton Transport Through One-Atom-Thick Crystals. *Nature* **2014,** *516* (7530), 227–230.

47. Duan, X.; Wang, C.; Pan, A.; Yu, R.; Duan, X. Two-Dimensional Transition Metal Dichalcogenides as Atomically Thin Semiconductors: Opportunities and Challenges. *Chem. Soc. Rev.* **2015,** *44* (24), 8859–8876.

48. White, R. M.; Lucovsky, G. Chemical Bonding and Structure in Layered Transition Metal Dichalcogenides. *Solid State Commun.* **1972,** *11* (10), 1369–1373.

49. Mak, K. F.; Lee, C.; Hone, J.; Shan, J.; Heinz, T. F. Atomically Thin MoS2: A New Direct-Gap Semiconductor. *Phys. Rev. Lett.* **2010,** *105* (13), 136805.

50. a)Mashtalir, O.; Naguib, M.; Dyatkin, B.; Gogotsi, Y.; Barsoum, M. W. Kinetics of Aluminum Extraction from Ti3AlC2 in Hydrofluoric Acid. *Mater. Chem. Phys.* **2013,** *139* (1), 147–152.b)Li, L. Lattice Dynamics and Electronic Structures of Ti3C2O2

and Mo2TiC2O2 (MXenes): The Effect of Mo Substitution. *Comput. Mater. Sci.* **2016,** *124*, 8−14.c)Sim, E. S.; Yi, G. S.; Je, M.; Lee, Y.; Chung, Y.-C. Understanding the Anchoring Behavior of Titanium Carbide-Based MXenes Depending on the Functional Group in LiS Batteries: A Density Functional Theory Study. *J. Power Sour.* **2017,** *342*, 64−69.

51. Anasori, B.; Lukatskaya, M. R.; Gogotsi, Y. 2D Metal Carbides and Nitrides (MXenes) for Energy Storage. *Nat. Rev. Mater.* **2017,** *2*, 16098.

52. Li, L.; Yu, Y.; Ye, G. J.; Ge, Q.; Ou, X.; Wu, H.; Feng, D.; Chen, X. H.; Zhang, Y. Black Phosphorus Field-Effect Transistors. *Nat. Nano.* **2014,** *9* (5), 372−377.

53. Roberts, K. Five Reasons Phosphorene may be a New Wonder Material. <https://nationalmaglab.org/news-events/feature-stories/phosphorene-wonder-material. >

54. Brent, J. R.; Savjani, N.; Lewis, E. A.; Haigh, S. J.; Lewis, D. J.; O'Brien, P. Production of Few-Layer Phosphorene by Liquid Exfoliation of Black Phosphorus. *Chem. Commun.* **2014,** *50* (87), 13338−13341.

55. Carvalho, A.; Wang, M.; Zhu, X.; Rodin, A. S.; Su, H.; Castro Neto, A. H. Phosphorene: From Theory to Applications. *Nat. Rev. Mater.* **2016,** *1*, 16061.

56. Mannix, A. J.; Zhou, X.-F.; Kiraly, B.; Wood, J. D.; Alducin, D.; Myers, B. D.; Liu, X.; Fisher, B. L.; Santiago, U.; Guest, J. R., et al. Synthesis of Borophenes: Anisotropic, Two-Dimensional Boron Polymorphs. *Science* **2015,** *350* (6267), 1513−1516.

57. Ji, J.; Song, X.; Liu, J.; Yan, Z.; Huo, C.; Zhang, S.; Su, M.; Liao, L.; Wang, W.; Ni, Z., et al. Two-Dimensional Antimonene Single Crystals Grown by van der Waals Epitaxy. *Nat. Commun.* **2016,** *7*, 13352.

58. Wang, H.; Yuan, H.; Sae Hong, S.; Li, Y.; Cui, Y. Physical and Chemical Tuning of Two-Dimensional Transition Metal Dichalcogenides. *Chem. Soc. Rev.* **2015,** *44* (9), 2664−2680.

59. Nicolosi, V.; Chhowalla, M.; Kanatzidis, M. G.; Strano, M. S.; Coleman, J. N. Liquid Exfoliation of Layered Materials. . *Science* **2013,** *340* (6139), 1226419−1226437.

60. Ferrari, A. C.; Bonaccorso, F.; Fal'ko, V.; Novoselov, K. S.; Roche, S.; Boggild, P., et al. Science and Technology Roadmap for Graphene, Related Two-Dimensional Crystals, and Hybrid Systems. *Nanoscale* **2015,** *7* (11), 4598−4810.

61. Kim, J.; Kwon, S.; Cho, D.-H.; Kang, B.; Kwon, H.; Kim, Y.; Park, S., et al. Direct Exfoliation and Dispersion of Two-Dimensional Materials in Pure Water via Temperature Control. *Nat. Commun.* **2015,** *6*, 8294.

62. Chung, C.; Kim, Y.-K.; Shin, D.; Ryoo, S.-R.; Hong, B. H.; Min, D.-H. Biomedical Applications of Graphene and Graphene Oxide. *Acc. Chem. Res.* **2013,** *46* (10), 2211−2224.

63. Brent, J. R.; Ganguli, A. K.; Kumar, V.; Lewis, D. J.; McNaughter, P. D.; O'Brien, P.; Sabherwal, P.; Tedstone, A. A. On the Stability of Surfactant-Stabilised Few-Layer Black Phosphorus in Aqueous Media. *RSC Adv.* **2016,** *6* (90), 86955−86958.

Synthetic routes to graphene preparation from the perspectives of possible biological applications

2

CHAPTER OUTLINE

2.1 Summary 17
2.2 Introduction 18
2.3 Bottom-Up Approaches 19
    2.3.1 Chemical Vapor Deposition Method 20
    2.3.2 Chemical Synthesis Method 21
2.4 Top-Down Approaches 23
    2.4.1 Mechanical Peeling 23
    2.4.2 Liquid-Phase Exfoliation of Graphene 23
    2.4.3 Exfoliating Agents 27
2.5 Conclusions 38
References 40

2.1 SUMMARY

Preparation of graphene by specific routes is described and their relevance to biological applications are compared in this chapter. After sufficient background discussion on numerous methods of graphene preparation, the major thrust of this chapter is devoted to the preparation of high-quality graphene for biological and biomedical applications. We exclusively discuss the production of unoxidized graphene with minor defects rather than graphene oxide or reduced graphene oxide, which are produced by vigorous chemical oxidation/reduction methods. The bottom-up synthesis of graphene using chemical vapor deposition (CVD) or chemical synthesis is briefly discussed, followed by more elaborate descriptions of liquid-phase exfoliation methods. The focus is then further narrowed down to biologically relevant, aqueous-phase exfoliation methods by ultrasonication, shear/turbulence, ball milling, and electrochemical routes. The efficiencies of these particular methods are compared in terms of graphene suspension concentrations (in mg/mL) produced per hour, with the highest value reported so far being 4 mg/mL/hour in water. The production cost of graphene by these methods is generally as low as $3—$5 per gram in water, highly economical compared to

Introduction to Graphene. DOI: http://dx.doi.org/10.1016/B978-0-12-813182-4.00002-7

commercially available graphene ($80–$100 per gram) or graphene oxide ($100–$120 per gram). Moreover, the possibility of using in expensive and common equipment such as an ultrasonicator and kitchen blender, without toxic byproducts as noted in graphene oxide synthesis, ensures the widespread use of these graphene suspensions. The resulting graphene layers were almost defect-free with I_D/I_G as low as 0.1–0.3 (ratio increases with defects), whereas the CVD method generally shows that of 0–0.1 ('defect free'). Furthermore, we discuss new opportunities that are available to make bio-grade, sterile graphene suspensions in cell media, sera or common biological buffers which are interesting and relevant to the biological community. These are water dispersible, stable, nearly defect-free graphene rather than commonly used alternatives such as graphene oxide or reduce graphene oxide, which are known to have numerous defect sites.

2.2 INTRODUCTION

Graphene applications in chemistry and biology are currently limited due to the limited availability of sufficient quantities of high-quality graphene. The preparation of high-quality graphene by inexpensive and scalable routes on a large scale is therefore an unmet urgent challenge. This chapter focuses on graphene preparation by bottom-up and top-down approaches, which are described later in this chapter (Fig. 2.1).[1] In the bottom-up method, briefly, the high thermodynamic stability of graphite over other carbon allotropes is utilized as a driving force to prepare graphene from molecular precursors. Multistep chemical synthesis of

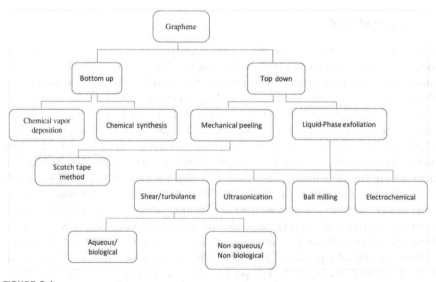

FIGURE 2.1

Major synthetic routes for graphene preparation.

graphene using synthetic organic methodology is also explored to make atomically precise graphene molecules of well-defined size and shape.[2] However, these methods are potentially promising for large-scale production of graphene but are currently limited to preparing only small quantities.

In the top-down method, on the other hand, the individual graphene layers are delaminated from graphite crystals by a variety of mechanical methods (Fig. 2.1). For example, the Scotch tape method was originally used to mechanically cleave graphite crystals to produce sufficient quantities of the single-layer graphene material.[3] This discovery led to the award of the Nobel Prize in Physics in 2010 and rapidly catalyzed the discovery of numerous fundamental and exciting properties of graphene. However, the Scotch tape method is quite limited in producing sufficient quantities of graphene for a large number of applications, such as for biological applications in drug delivery, biosensing, or theranostics.

Alternatively, top-down methods are elaborated in this chapter to demonstrate their potential for scale-up of high-quality graphene production in aqueous solvents or biological media such as aqueous media, cell growth media or even serum at a low cost. Thus, the latter methods are quite attractive to produce biological-grade, sterile, graphene—water suspensions for a variety of biological applications. Such preparations may be termed "biographene" due to their compatibility with biological systems such as cells, oragnelles, or even nematodes, as already demonstrated in our laboratory.[4−6]

Separation of graphene layers from the crystal by micromechanical forces in a liquid phase, top-down method, is commonly referred to as exfoliation.[7] Specific stabilizing agents or solvent media are often required for exfoliation, and the exfoliated sheets tend to aggregate in the absence of such a stabilizing agent, although many exceptions to this have been noted.[7,8] The micromechanical delamination and fragmentation of graphite crystals is carried out using ultrasonication, shear/turbulent force, ball milling, or electrochemical potential. Exfoliation can be performed in organic solvents or in water. Aqueous-phase exfoliation of graphite generally requires an exfoliating agent, which can be biologically benign or not.

The major focus of this chapter, however, is graphite exfoliation under biologically relevant conditions, in aqueous media, for applications in chemistry and biology. First, a general survey of synthetic strategies for the preparation of graphene will be described, and this will be followed by a detailed description of aqueous-phase exfoliation methods. The above methods of graphene preparation or synthesis are compared here in terms of their yield, purity, scalability, and relevance to biological applications.

2.3 BOTTOM-UP APPROACHES

In the bottom-up strategy, graphene is precisely prepared by the decomposition of molecular precursors as in CVD or built from benzene-like synthons by standard synthetic organic chemistry methodology. Graphene made from multistep organic synthesis is referred to as "molecular graphene." Both of these bottom-up approaches result in atomically precise single-layer graphene sheets, which are highly desired for electronic applications. While CVD is used to make large flakes of graphene, the synthetic molecular graphene is quite small in size (a few nm in diameter) and preparing micron-size flakes is a current, unmet challenge. However, these methods are not suitable for applications that need large quantities of inexpensive, high-quality graphene.

2.3.1 CHEMICAL VAPOR DEPOSITION METHOD

In the solid-phase methods, which are also referred to as dry methods in Fig. 2.1, single-layer graphene is prepared from molecular precursors on a supporting substrate surface of metals and semimetals.[9] Graphene films are grown by the decomposition of specific carbon sources,[10] and CVD is carried out under a variety of experimental conditions such as high-temperature, plasma-enabled, low-pressure, high-pressure, and so on, depending on the substrate, precursors, and more importantly the target application. Carbon atoms are formed under CVD conditions and nucleated on to the substrate surface to form graphene, under vacuum.[10] When methane is used as the carbon source (Fig. 2.2), e.g., the following reaction occurs to form graphene:

$$CH_4 \rightarrow C(graphene) + 2H_2 \tag{2.1}$$

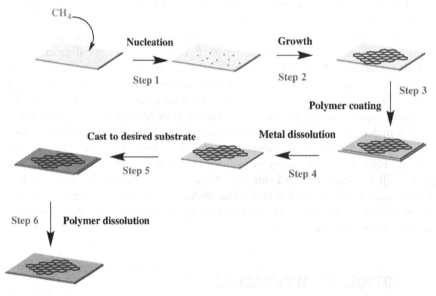

FIGURE 2.2

Chemical vapor deposition (CVD) of methane on metal surface to from graphene, followed by casting onto the desired surface using polymer-based templates.

Thermodynamically, the above reaction is not feasible at room temperature but favorable at $>650°C$ (theoretical estimate).[10] Deposition of defect-free graphene from methane by CVD is also entropically unfavorable under normal conditions. However, the process is favored with substrates which have high lattice match with graphene and also at low graphene to metal solubility at high temperatures.[10] Catalytic decomposition of carbon by active sites on the metal surface facilitates the growth of graphene sheets. The CVD method was used to prepare defect-free graphene layers as big as $>1 \text{ cm}^2$, but large-size preparations are time consuming, expensive, not readily scalable. Further improvements are needed to address these issues.[9,10]

Graphene grown on metal surfaces such as Cu or Ni by CVD is not convenient for electronic applications, especially for use in flexible and transparent devices. Transfer of graphene layers from these metals onto a polymer or semiconducting surfaces is needed for such applications, and the latter supports are not directly usable in CVD due to their poor stability to high temperatures or plasma.[11] Transfer of graphene from metal substrates is carried out using polymethylmethacrylate (PMMA) as a sacrificial polymer, for example.[11] Typically, the PMMA is coated on to a graphene/substrate (Fig. 2.2, Step 3), followed by removal of the metal substrate by acid etching (Fig. 2.2, Step 4). Then, the PMMA/graphene membrane is cast onto the target substrate, followed by degradation of PMMA either by using high temperature $(300-500°C)$ or dissolution by exposure to organic solvents (Fig. 2.2, Step 6). The transfer is further facilitated by applying it to the roll-to-roll process.[9] However, this method of using polymer supports to transfer the graphene can leave traces of the polymer residue and/or metallic residue on graphene, and these impurities are undesirable or deleterious for many high-end electronic or other sensitive applications.[11]

CVD, when combined with the transfer approaches, was successful to produce atomically precise, high-quality graphene, which is suitable for many applications in the electronics industry. However, the low yield and high cost of this method limit its application. Biological applications of CVD-derived graphene, e.g., are severely limited for these reasons.[12] Since CVD is a major subject of discussion by itself, we recommend interested readers to books or review articles devoted for this approach (see Refs. 10 and 11).

2.3.2 CHEMICAL SYNTHESIS METHOD

Precisely sized and shaped nanographene or nanoribbons are made from small organic molecules using multistep chemical synthesis or by polymerization of suitable monomers.[2] An example of graphene synthesis is shown in Fig. 2.3, where fused aromatic rings and polyaromatic compounds undergo a series of reactions to form graphene nanoribbon. These high aromatic compounds (>200 carbon atoms) are poorly soluble in solvents, and hence, substitution of long-chain alkyl groups (e.g., C12) at specific positions is used to improve the solubility in organic solvents. Graphene nanoribbons with variable length was controlled

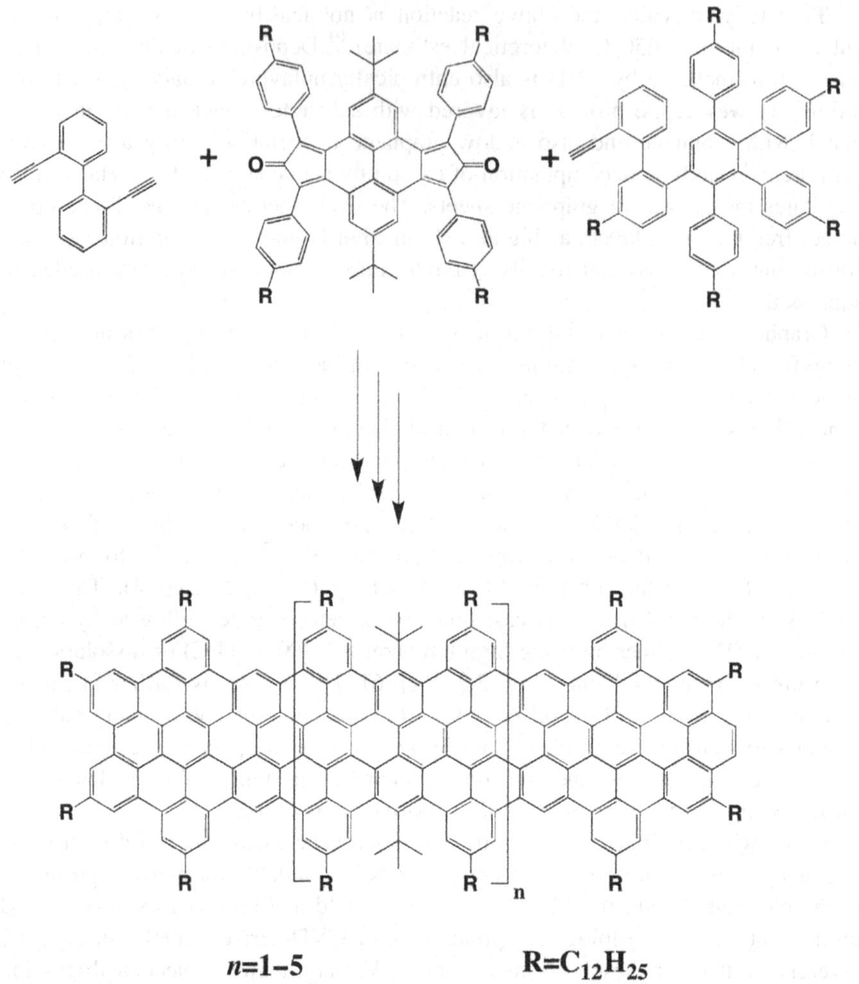

FIGURE 2.3

Chemical synthesis of precisely constructed graphene sheets.[2]

by varying the stoichiometric ratio at the cycloaddition reactions.[13] By this way, a family of five molecules was synthesized with 132 to 372 total carbons in a single molecule, which are characterized using MALDI-TOF and absorption spectroscopy. However, higher oligomers ($\sim$350 carbons) were insoluble in organic solvents even with the alkyl chain substitutions.[2]

Chemical synthesis, in contrast to CVD synthesis, results in graphene suspensions in organic liquids, which can cast directly on to desired substrates show promising performance in organic transistors.[13] However, the time and cost required for large-scale synthesis of graphene by this method limit its application, and it is not considered a viable alternative at this stage. Also, it should be noted that the large-area graphene molecules are synthesized by chemical routes lack solubility, which restrains the complete purification and characterization of these molecules.[13]

2.4 TOP-DOWN APPROACHES

Graphene can be isolated from multilayer graphite crystals by various chemical and mechanical routes.[9] Since graphite is an inexpensive and abundant resource, these methods have paramount importance in using graphene for large-scale applications.[7] Compared to graphene prepared by CVD or chemical synthesis, graphene derived from graphite exfoliations is highly polydispersed and has few-layer structures. The solution processability of exfoliated graphene is attractive to chemists and biologists for various applications.[9]

2.4.1 MECHANICAL PEELING

The first successful isolation of graphene from graphite crystals was accomplished by mechanical peeling of graphite using Scotch tape, which led to the Nobel prize for physics in 2010.[3] In this method, the glue in the Scotch tape strongly binds to the surface of graphite and separates the top layers from the large crystal. Repeated peeling of the few layers of graphene stuck to the Scotch tape eventually results in atomically thin sheets. However, this method lacks scalability but research on electronic nature of graphene is still studied by adopting this method.[9]

2.4.2 LIQUID-PHASE EXFOLIATION OF GRAPHENE

From the perspectives of the dry methods described above, liquid-phase exfoliation of graphene is advantageous to make large-scale, highly concentrated dispersions in numerous solvent systems.[7] Delamination of graphite to graphene layers assisted by a solvent or molecules called exfoliating agents is accomplished here. Stable graphene dispersions can be achieved after separating the unexfoliated graphite by centrifugation, without any chemical processing (Fig. 2.4). Two primary components are essential for this process:[7] (1) the mechanical force required to overcome the interlayer interactions between adjacent layers in the graphite

FIGURE 2.4

Exfoliation of graphite to graphene.

crystal and (2) a stabilization agent which stabilizes the graphene flakes and prevents their restacking to form thermodynamically more stable graphite crystal. The exfoliation can be achieved by ultrasonication, high shear mixing, ball milling, or electrochemical potential.

2.4.2.1 Ultrasonication

Based on variations of the method used to produce the necessary mechanical force to delaminate the graphite crystal and the stabilizer, several exfoliation methods are known to date and new methods are continuously being invented. One of the commonly used mechanical methods to exfoliate graphite is ultrasonication.[14] Ultrasonic waves produce high-energy (equivalent to ∼5000 K) microbubbles, which induce delamination and fragmentation of graphite sheets.[15] The ultrasound is generated by either bath sonicator or by tip sonicators at a controlled temperature. Under suitable conditions, few-layer graphene, ranging from 1 to 10 layers, is easily produced and the method is scalable to produce large amounts, albeit slowly.[14] On average, sonication mediated exfoliation methods displays exfoliation rates about 0.05−0.7 mg/mL/h in a variety of organic or aqueous solvents. Some examples of ultrasonication-induced graphite exfoliation in organic solvents or surfactants are given in Table 2.1. The ultrasonication method also produced graphene in biologically relevant conditions, such as in proteins or DNA, as shown in Table 2.6. A complete literature survey of sonication-assisted graphite exfoliation is reviewed in Ref. 18.

2.4.2.2 Ball milling

Ball milling of graphite with appropriate stabilizers is another mode of exfoliation in liquid phase.[21] Graphite is ground under high sheer rates with millimeter-sized metal balls causing exfoliation to graphene (Fig. 2.5), under wet or dry conditions. For instance, this method can be employed to produce nearly 50 g of graphene in the absence of any oxidant.[22] Graphite (50 g) was ground in the ball mill with oxalic acid (20 g) in this method for 20 hours, but, the separation of

Table 2.1 Representative Examples of Liquid-Phase Exfoliation of Graphite Using Ultrasonication

Medium of Exfoliation	Rate of Exfoliation (mg/mL/h)	Initial Graphite Concentration (mg/mL)	Final Graphene Concentration (mg/mL)
NMP[16]	0.3	3.3	1.2
DMF[17,18]	0.04	5	0.06
Sodium cholate in water[19]	0.013	5	0.3
Pyrene sulfonic acid in water[20]	0.7	20	0.7
Sodium dodecylbenzene sulfonic acid (SDBS) in water[20]	0.22	50	0.22

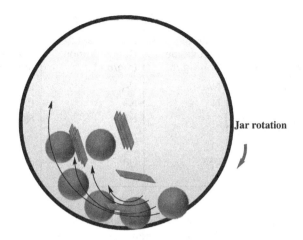

FIGURE 2.5

Ball milling of graphite to graphene in a rotating jar.

Table 2.2 Examples of Liquid-Phase Exfoliation of Graphite Using Ball Milling

Medium of Exfoliation	Milling Time (h)	Initial Graphite (g)	Amount of Graphene Produced (g)
Oxalic acid[22]	30	50	50
KOH[23]	8	0.2	NA
Ammonia borane[24]	4	0.3	~0.3
1-Pyrenecarboxylic acid/methanol[26]	30	0.1	NA
Sodium chloride[27]	1	0.05	~0.037
Melamine[28]	0.5	0.005	~0.0024

unexfoliated fraction was not discussed.[22] Similarly, solvent-free graphite exfoliations were carried out under dry milling conditions using KOH,[23] ammonia borane,[24] and so on. The list of graphite exfoliations performed using ball milling is given in Table 2.2. However, the metallic impurities from the machinery used for ball milling are a major disadvantage of this method for certain applications.[25]

2.4.2.3 Electrochemical exfoliation

Recently, electrochemical exfoliation of graphite was explored to produce high concentrations of graphene dispersions in the liquid phase.[29] In this method, a suitable potential is applied across a graphite rod, and oppositely charged exfoliating agents are allowed to intercalate into the graphite layers, causing interlayer expansion. Depending on the charge on the working electrode and the presence of specific anions (e.g., SO_4^{2-}) or cations (e.g., Li^+), the method works rapidly.

Table 2.3 Examples of Electrochemical Exfoliation of Graphite

Medium of Exfoliation	Rate of Exfoliation (mg/mL/h)	Initial Graphite Concentration (mg/mL)	Final Graphene Concentration (mg/mL)
0.1 M Ammonium sulfate[30]	NA	NA	2.5 (DMF)
Sulfuric acid[31]	NA	NA	0.085 (DMF)
1-Butyl-3-methylimidazolium tetrafluoroborate [BMIm] [BF$_4$] in water[32]	NA	NA	NA
LiCLO$_4$ and tert-butyl ammonium cation[33]	0.007	NA	.027

The expanded graphite requires further delamination by sonication or shear to from few-layer dispersions. This route is highly promising and achieved fast exfoliation of graphite (<10 minutes, but the initial concentrations of graphite were not reported in the studies for a fair comparison) (Table 2.3).[30,31] Also, the electrochemically intercalated graphite was further exfoliated using ultrasonication in suitable organic solvents.[29] Other methods of graphene preparation such as high jet flow, supercritical carbon dioxide, and vortex flow shearing have also been used to date.[15] However, limited work has been done to evaluate the relative success of these methods.

2.4.2.4 Shear exfoliation

Another common method to generate a high mechanical force to separate the graphene layers from the graphite crystal is by applying high shear. Normal force parallel to the plane of graphene sheets in graphite separates individual layers by overcoming the interlayer interactions. In another pathway, small graphite platelets are formed by the rapid collisions between (larger) graphite platelets. This facilitates the delamination process because small-area sheets possess a small contact area, and thereby lower interlayer interactions (Fig. 2.6). These mechanisms show high exfoliation efficiencies compared to ultrasonication. High shear dispersers have been used to rapidly exfoliate graphite in high yield.

The shear force required to delaminate graphite (rate $\sim 10^4 \text{ s}^{-1}$)[34] is generated using a rotor and stator assembly.[34] In suitable solvent systems, few-layer graphene dispersions of 2–5 layers are routinely produced in large volumes, as large as 300 L (in 10% sodium cholate).[34] Interestingly, the shear force coupled with turbulence produced in a regular kitchen blender (rate $\sim 10^6 \text{ s}^{-1}$) was able to exfoliate graphite very rapidly.[4,35] This may be an inexpensive and widely available method for many laboratories globally. A wide range of exfoliating agents and stabilizing systems were used to produce high-concentration, near defect-free graphene suspensions (Table 2.4). Biological exfoliating agents used under shear conditions are listed in Table 2.4. The shear force achieved in the disperser or a kitchen blender causes delamination of graphite by normal force parallel to the graphite plane and

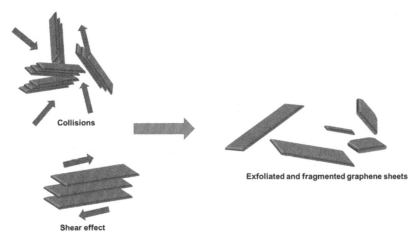

FIGURE 2.6

Mechanism of shear- and turbulence-based exfoliation of graphite.

Table 2.4 Examples of Liquid-Phase Exfoliation of Graphite Under High-Speed Shearing

Medium of Exfoliation	Rate of Exfoliation (mg/mL/h)	Initial Graphite Concentration (mg/mL)	Final Graphene Concentration (mg/mL)
NMP[24]	1.3×10^{-3}	50	0.1
DMF[36]	0.03	3	0.22
Sodium cholate in water[34]	0.25	100	1.0
2% Polyacrylic acid in water at pH 3.5 (H_2SO_4)[37]	5	20	2

collisions between graphite platelets enhances plate fragmentation (Fig. 2.6).[4] More on this method is presented under the biological exfoliation methods. The yields of shear-based exfoliations are much higher than ultrasonication methods and about 1−4 mg/mL/hour exfoliation efficiency was achieved by this method.

2.4.2.5 Comparison of common exfoliation methods
The exfoliation methods listed above have their own advantages and disadvantages as tabulated here (Table 2.5). The major emphasis in this comparison is on the ease of experiment, cost, scalability, availability, and exfoliation efficiency.

2.4.3 EXFOLIATING AGENTS

An exfoliating agent is a molecule which assists the delamination of graphene and also prevents the reaggregation of graphene layers, described in the above

Table 2.5 Comparison of Common Exfoliation Methods

Method	Advantages	Disadvantages
Ultrasonication[14]		
	Simple apparatus	Low exfoliation efficiency
	Available in most labs	Low scalability
	Smaller volumes (in μL) can be used	Poor product/wattage ratio
Shear/Turbulence[15,4]		
	High exfoliation efficiency	Not a common apparatus
	Scalable (300 L is reported)	Working volume is 10 mL for common instruments
	Simple instrumentation	High initial graphite concentration is required
	An inexpensive kitchen blender is an alternative[4,34]	Frothing issues with some exfoliation agents
Ball Milling[20]		
	Good exfoliation efficiency	High chances of metal contamination
	Scalable to gram-scale production	Not a common apparatus
	Cheap instrumentation	
Electrochemical[29]		
	High production rate	Large-area graphite crystals or rods may be needed
	Possibility of exfoliation and functionalization, simultaneously	Poor exfoliation: other mechanical forces for delamination
	Inexpensive	High chances of unwanted oxidation

studies. The exfoliating agent can be either solvent molecules (e.g., *N*-methylpyr-rolidone, NMP) or a third component. To stabilize graphene in solution, the exfoliation agent should have stronger interactions with graphene sheets, than graphene—graphene interlayer interactions in the graphite crystal.[38] Thus, the graphene layers stabilized in solvent are a graphene/exfoliating agent complex. This can be compared to a simple solvation model where graphite (G_n) is solvated in a solvent or bound to another molecule (S).

$$G_n + mS \rightarrow (G_x S_y), \text{ where } n, \ m, x \text{ and } y \text{ are integers}$$

The magnitudes of x and y are determined by the efficiency of exfoliation and graphene/exfoliating agent interactions. A very good mode of exfoliation should result in lower values of x ($x = 1$ is monolayer graphene) and this value is dependent on the ability of exfoliating agents to promote and protect the formation of thin graphene layers.[7]

From a biological perspective, the medium of exfoliation is the most important consideration to justify whether a graphene sample is valuable or not. In other words, the exfoliating agent used for preparation of graphene should be benign to biological systems. Hence, we divide the exfoliation media used for graphene into biological and nonbiological for further discussion.

2.4.3.1 Nonbiological exfoliating agents

Liquid-phase exfoliation of graphite was initially demonstrated using organic solvents or surfactants as exfoliating/stabilizing agents. N-methyl-2-pyrrolidone (NMP) and dimethyl formamide (DMF) was the widely used solvent system for this approach.[18] The solvent parameters and ability to form stable graphene/solvent complexes made the exfoliation feasible in these solvents. Similarly, the amphiphilic nature of surfactants was utilized in surfactant-based exfoliations. The detailed mechanism of exfoliation using these systems is discussed further below.

2.4.3.1.1 Organic solvents as exfoliating agents:

Pure organic solvents such as NMP, DMF, benzyl benzoate, and γ-butyrolactone show the ability to exfoliate graphite by balanced graphene solvent interaction and coulombic repulsions. Solvents with matching surface energies of graphite ($70-80$ mJ/m^2) are found to be successful in exfoliating graphite.[40] Solvents with 40 mJ/m^2 would be able to separate the graphite layers effectively by minimizing the interlayer interactions. Thus, solvents matching this surface energy, NMP (40 mJ/m^2) and DMF (37.1 mJ/m^2), demonstrate good exfoliation abilities with nearly 30% single-layer graphene sheets in dispersion. The thermodynamic relation of mixing of graphite in a solvent is approximately:[38]

$$\frac{H_{mix}}{V_{mix}} = \frac{2}{T_{NS}}\left(\sqrt{E_{SS}} - \sqrt{E_{SG}}\right)^2 \varnothing_G \qquad (2.2)$$

where H_{mix} and V_{mix} are the enthalpy and volume of mixing, respectively, T_{NS} is the thickness of the graphene sheet, and E_{SS} and E_{SG} are the surface energy of solvent and graphene, $\varnothing_G$ is the volume fraction of dispersed graphene, respectively.

Other fair solvents such as *ortho*-dicholorobenzene (*o*-DCB), 3,3′-iminobis (N, N-dimethylpropylamine) (DMPA), N-[3-(dimethylamino)propyl]methacrylamide (DMAPMA), 2-(tert-butylamino) ethyl methacrylate (BAEMA), 2-(dimethylamino) ethyl methacrylate (MAEMA), and chlorosulfuric acid show good exfoliation rates (Fig. 2.7), but the cost, high boiling point, and toxicity (NMP showed reproductive toxicity to humans) of these solvents are major concerns for industrial uses.[38]

Exfoliation rates about 20 μg/mL/hour are achieved using these solvents by ultrasonication of graphite in these solvents.[14] The exfoliation efficiencies are improved by mixed-solvent systems and other additives to the medium. For example, *n*-octylbenzene enhanced the exfoliation rates up to five-fold in NMP (15% by volume of *n*-octylbenzene), possibly by high $\pi - \pi$ interaction of the aromatic octylbenzene.[39] In support, naphthalene, another aromatic system, also enhanced

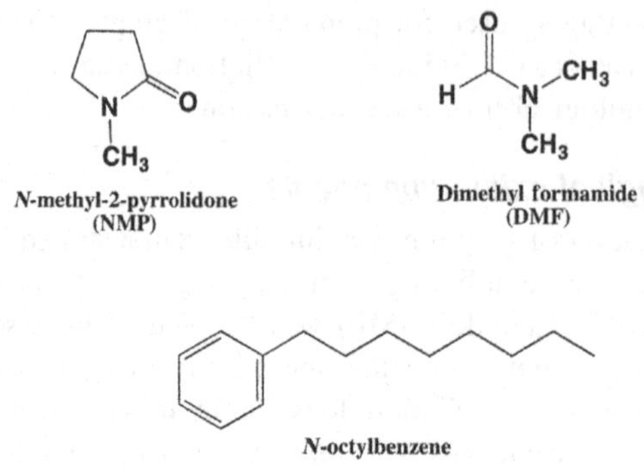

FIGURE 2.7

Chemical structure of organic molecules used for graphite exfoliation.

(nearly twofold) the yield of pristine graphene exfoliated in NMP, which is operated via $\pi - \pi$ interactions.[17] However, the role of these molecule in increasing or decreasing the surface energy of the solvents has not yet been investigated.

2.4.3.1.2 Surfactants as exfoliating agents:

The high surface energy of water (72.8 mJ/m^2) is unfavorable for exfoliation of graphite.[39] Surfactants are used to lower the surface energy of water to that of graphene and either by sonication or shearing are used to delaminate graphite (Fig. 2.8). The first reported surfactant used for graphite exfoliation was sodium dodecylbenzene sulfonate (SDBS)[40] and, today, hundreds of molecules that can exfoliate graphite in water are known.[40] One major class of these surfactants is aromatic, particularly consisting of polyaromatic hydrocarbons. Pyrene derivatives have been widely used for exfoliation of graphite, perhaps because of their commercial availability, chemical reactivity, and large π surface. For example, 1,3,6,8-pyrenetetrasulfonic acid, has strong affinity to graphene basal plane and the polar sulfonic acid groups protect from reaggregation of graphene via repulsive interactions.[40] N,N'-dimethyl-2,9-diazaperopyrenium dication (MP^{2+}2Cl$^-$) is an example of a cationic surfactant which showed exfoliation in water upon ultrasonication for 30 minutes.[41] However, the pyrene derivative of MP^{2+}, with smaller π surface, did not show exfoliation, even after prolonged sonication (24 hours), emphasizing the role of high $\pi - \pi$ interaction of the exfoliating agent with graphite.

Nonionic aromatic surfactants were also used for exfoliation of graphite in aqueous and nonaqueous conditions. Perylenebisimide bolaamphiphile was used to prepare few-layer graphene in biologically relevant, phosphate buffer at pH 7.4.[42] Other exfoliating agents in this class include anthracene carboxylic acid (in an ethanol/water mixture), porphyrins, phthalocyanines, and many other pyrene derivatives.[39] Most of these exfoliating agents need to be prepared by multistep synthetic methods that hamper the commercial production of graphene using these

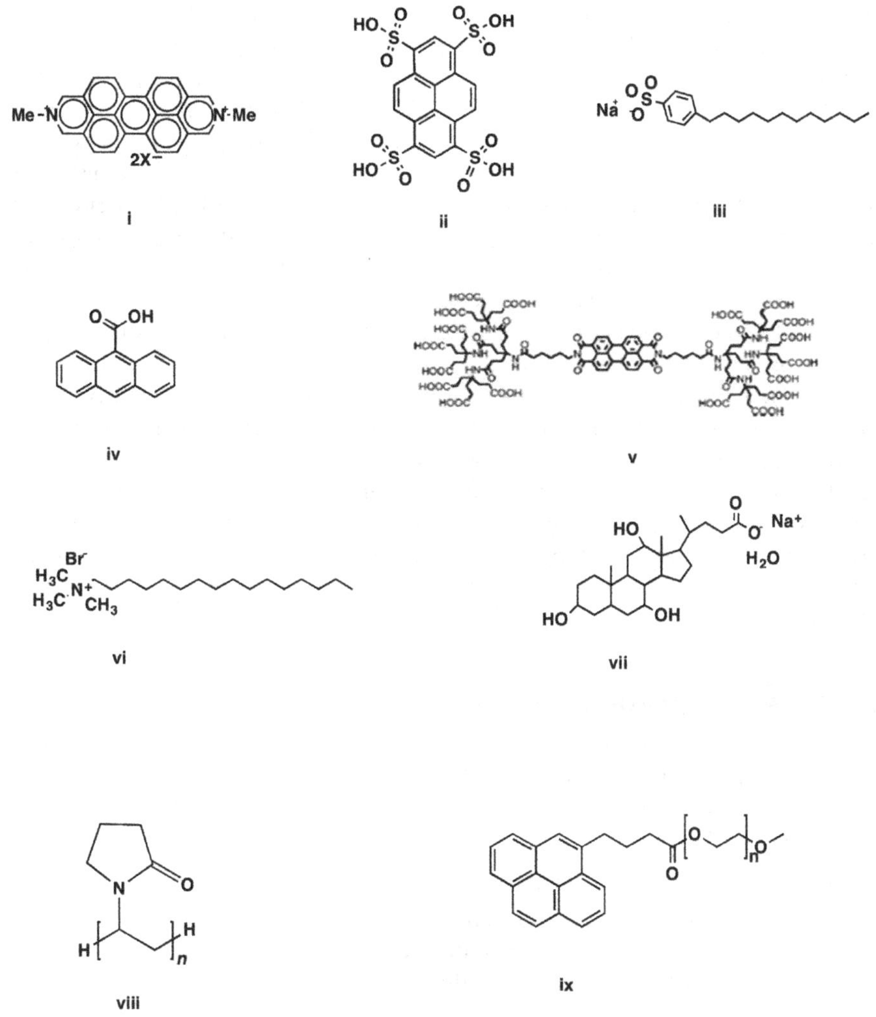

FIGURE 2.8

Chemical structures of few surfactant molecules used for graphite exfoliation.

reagents. Furthermore, most of the polyatomic compounds are mutagens and/or genotoxic substances, which cannot be used in biological applications.[43]

Nonaromatic surfactants with long aliphatic chains were demonstrated to exfoliate graphite in solution. Cetryltrimethyl ammonium bromide, sodium cholate, TritonX-100, arachidic acid, and nanocellulose are a few examples of this class.[44] Sodium cholate, a bile salt, showed high exfoliation efficiency using both ultrasonication and shear methods, and is a widely studied molecule for graphite exfoliation. An exfoliation rate of about 1 mg/mL/hour was achieved.[14]

Polymeric surfactants with aromatic side groups and hydrophilic groups were also designed for graphite exfoliation in water. Polyvinylpyrrolidone was first employed to exfoliate graphite, not only because it is a water-soluble polymer but also because of its structural similarity to NMP.[45] Polymers of pyrene, which is another exfoliating agent of graphene, were also synthesized for aqueous exfoliations. Pyrene-polyethylene glycol and pyrene-polycaprolactone were used to delaminate graphite using supercritical carbon dioxide in DMSO (50% yield w/w) and in water (10% yield w/w).[38] Such exfoliation methods followed by composite preparation of graphene/polymer is useful for many electronic and material applications and will be discussed in a separate chapter.

Exfoliation of graphene using organic solvents or surfactants showed various degrees of success, but the lack of a commercial or practical use is reflected in many reports. Strong or prolonged sonication conditions can cause extensive fragmentation of the sheets that affects the electronic properties of graphene in a device. Separating graphene from the exfoliating agent or solvent for device applications in film state is extremely difficult and most leave residues. For example, NMP forms its polymer under sonication conditions and strongly binds to graphene sheets. Another important concern is the toxicity of the molecules used for exfoliation, which makes it untenable for commercialization and also for biological research.

2.4.3.2 Biological exfoliating agents

Exfoliating agents derived from biological systems and molecules benign to biological organisms can be termed as biological exfoliating agents. A few key questions to consider before using an exfoliating agent for graphene for a biological experiment would be:

1. Is it compatible with the experimental system under investigation?
2. Can the molecule exfoliate and stabilize graphene in the media?
3. Would this molecule interfere with the bioassay or analytical methods?
4. What is the availability and cost of the exfoliating agent?

A wide variety of biomolecules were used to exfoliate graphite in water including proteins, carbohydrates, polysaccharides, and animal sera.[46] The advantage of using biomolecules for exfoliation is that the biocompatibility of the media is guaranteed. Biocompatible molecules such as polyethylene glycol, aliphatic acids, and different small organic molecules were also used for exfoliation of graphite, which may be suitable for bioassays.[38] Proteins without biological or catalytic activities, are suitable candidates for graphene preparation for biological applications. The cost of the protein should also be taken into consideration for large-scale applications. We have demonstrated that inexpensive albumins can exfoliate graphite under high shear or ultrasonication conditions.[4–6] The cell culture media generally have about 10% serum as a nutrient, and are also used to produce high concentrations of near defect-free graphene under sterile conditions.[6] However, the choice for exfoliation agent can be very specific for certain applications.

A comparison of biomolecule-assisted exfoliations is shown in Table 2.6. The rate of exfoliation (mg/mL/hour) is used here as a critical factor to compare different methods. The production rate (mg/hour) is used in the literature, but exfoliation volume varies by method. The percentage yield is another way of expressing the yield (wt%), but initial graphite concentrations are not the same. Moreover, graphite is a cheap resource and unexfoliated graphite is recyclable after separating from graphene. Thus, expressing efficiency in terms of concentration of graphene in the medium over time is more reasonable.

Many of the aforementioned molecules were able to exfoliate graphite, but the underlying mechanisms of exfoliation or stabilizing effects are poorly understood. The suggested mechanism in most of cases is that the hydrophobic sites of the molecule interact with graphene sheets and the hydrophilic sites give dispersion stability to graphene, which is comparable to the formation of micelles from amphiphilic molecules. A deep investigation to follow the mechanism of protein-based exfoliation of graphene revealed that only proteins with sufficient surface charge exfoliated or stabilized graphene in water.[4] This refutes the understanding that all amphiphilic molecules can exfoliate or stabilize graphite. However, the key question remains whether the protein facilitates the exfoliation or kinetically traps the graphene layers and inhibits the aggregation of the layers (Fig. 2.9).

Among various biological exfoliation agents, detailed studies have been carried out for protein-based methods to understand the mechanism of exfoliation, using shear and turbulence method.[4] Common and inexpensive proteins, BSA, ovalbumin, lysozyme, hemoglobin, and beta-lactoglobulin were used to examine the kinetics of exfoliation in water (Fig. 2.10). Exfoliation of graphite was not feasible with all the proteins, but proteins with substantial surface charge (> -15) showed exfoliation capability.[4] No other factors, such as hydrophobicity, amino acid constitution, or aliphatic contents showed direct trend towards the exfoliation efficiency of proteins. Exfoliation at varying pH (2−10) using BSA (isoelectric point of 4.2) resulted in maximum exfoliation rates above neutral pH, which corroborates the need for sufficient surface charge on the protein for graphite exfoliation.[4]

The exfoliation efficiency of proteins was also dependent on factors including $[graphite]_o$, $[BSA]_o$, volume, and speed of the blade. The rate of exfoliation by varying these factors is resolved in the following equation:[4]

$$\text{Rate of exfoliation} \quad \propto \quad [\text{BSA}]_o^0 \, [\text{Graphite}]_o^{1.2} \, (\text{pH})^{0.29} \, (\text{Volume})^{0.99} \, (\text{Blade speed})^{1.7}$$

$$(2.3)$$

The powers represent the order of the reaction for that factor. This equation is valid above 3 mg/mL $[BSA]_o$. The equation suggests that increasing the initial concentration of graphite and blade speed can still produce higher exfoliation rates. Factors such as viscosity of the medium, blade shape, and length could also affect the rate of exfoliation, which was similar in this case. Overall, the role of

Table 2.6 Biomolecules Used for Exfoliation of Graphite

Exfoliating Agent	Category	Method	Exfoliation Efficiency (mg/mL/h)	Comments
Bovine serum albumin (BSA)[47]	Protein	Ultrasonication	0.27	Unexfoliated graphite was separated by settling for 24 h
BSA4	Protein	Shear	4	Highest exfoliation rate in protein-based methods
Beta-lactoglobulin[4]	Protein	Shear	3.5	
Ovalbumin[4]	Protein	Shear	1.3	
Lysozyme[48]	Protein	Ultrasonication	0.35	Hybrid material showed anticancer activity
Hydrophobins[49]	Protein	Ultrasonication	0.75	60% ethanol in water is used for exfoliation
Gelatin[50]	Protein/peptide	Ultrasonication	0.04	Gelatin is not water soluble at room temperature
Calf histone[51]	Protein	Ultrasonication	NA	Yield is not reported
Animal sera[6]	Complex protein mixture	Shear	6	10% serum from six different animal sources
Melamine[51]	Small molecule	Ball milling	0.5	Melamine is potentially toxic (LD50$_{rat}$ -3.8 mg/kg)
Urea[52]	Small molecule	Ultrasonication	0.12	Urea is toxic (LD50$_{rat}$ -8.5 mg/kg)
Saccharin[53]	Small molecule	Electrochemical	5.7 (at 10V)	Oxidation or high-degree defects are noted in Raman spectrum
Amphotericin B[54]	Small molecule	Ball milling	0.45	Exfoliated in DMF and suspended in water
DNA[55]	Nucleotide	Ultrasonication	0.4	
RNA[56]	Nucleotide	Ultrasonication	NA	
Flavin mononucleotide[57]	Nucleotide	Ultrasonication	0.06	
Pullulan[58]	Polysaccharide	Ultrasonication	1	
Gum arabic[59]	Polysaccharide	Ultrasonication	0.006	
Cellulose[60]	Polysaccharide	Ultrasonication	0.25	Stable for 4 months
Lignin[61]	Polysaccharide	Ultrasonication	0.04	Alkali lignin was used
Chitosan[61]	Polysaccharide	Ultrasonication	11	Stable for 1 week

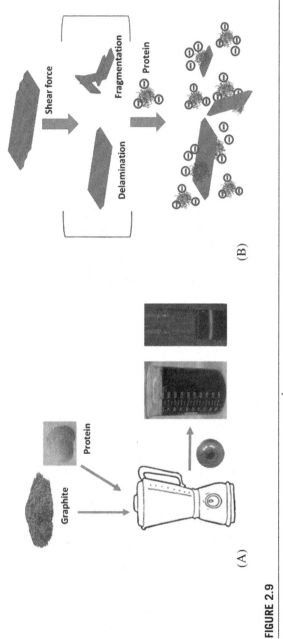

(A)

(B)

FIGURE 2.9

(A) Protein-based exfoliation in a kitchen blender[4] (copyright© John Wiley & Sons). (B) Proposed mechanism of protein-based exfoliation of graphite.

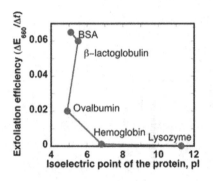

FIGURE 2.10

Exfoliation efficiency of different proteins as a function of their isoelectric point.[4]

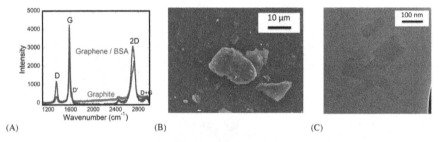

FIGURE 2.11

(A) Changes in Raman spectrum of graphite after exfoliation. (B) SEM image of graphite. (C) TEM image of exfoliated graphite.[4]

proteins in the exfoliation event can be deduced to (1) lower the surface tension of the medium, (2) trap the exfoliated graphene sheets and prevent them from restacking, and (3) provide enough surface charge to stabilize graphene in colloidal form.

Protein-exfoliated graphene was characterized using Raman spectroscopy and transmission electron microscopy to find the number of layers, lateral size and degree of defects using the equations below (2.3, 2.4). The major differences in Raman spectra of graphene and graphite are the evolution of D band (~ 1345 cm^{-1}), change in shape and position of 2D band (~ 2700 cm^{-1}), and appearance of D$'$ band (~ 1620 cm^{-1}), as a result of delamination (Fig. 2.11A)[34] (detailed characterization techniques and band assignments will be discussed in following chapters).

$$<N_G> = 10^{0.84M + 0.45M^2}; M = \frac{(I_{\omega_p}/I_{\omega_s})_{\text{Graphene}}}{(I_{\omega_p}/I_{\omega_s})_{\text{Graphite}}} \tag{2.4}$$

N_G, number of layers; I, peak intensity; ω_p, 2D peak (~ 2725 cm^{-1}); ω_s, 2D shoulder peak (~ 2695 cm^{-1}).[34]

$$<L> = \frac{k}{(I_D/I_G)_{G'\text{ene}} - (I_D/I_G)_{G'\text{ite}}} \tag{2.5}$$

L, lateral size of the sheets; k, 0.17 (experimentally calculated); I_D/I_G is the intensity ratio of D and G band.

When graphite (Fig. 2.11B) was exfoliated in a kitchen blender with BSA or animal sera, 3−5-layered graphene was produced, which had 600-nm lateral size, on average, characterized by both Raman spectroscopy and TEM images. TEM images were collected for about 100 sheets and the largest length of the sheet was measured using NIH/imageJ free software, after calibrating the image using the scale bar. The presence of few-layered graphene and the submicron lateral size of the sheets (Fig. 2.11C) was confirmed in TEM images.

Raman spectra calculations showed minor edge defects to graphene sheets due to shearing, and more importantly no observable oxidation occurred. The extent of defect was analyzed using Raman spectroscopy, where the D band and D′ band intensities represent defect activation (Fig. 2.11A).[34] Biographene produced here showed an $I_D/I_{D'}$ ratio of 1.7−2.7, representing the edge defect region. For comparison, chemically oxidized graphene oxide showed an $I_D/I_{D'}$ ratio of 7−13.[34]

The kinetic stability of graphene in biological media is a major concern in applying these materials for solution-processable methods or for long-term handling/shipping. The storage stability of these materials is often ignored but plays a crucial role in performance and toxicity assessments. Graphite exfoliated in BSA solutions showed a long shelf-life of over 2 months, dependent on the storage temperature (Fig. 2.12),[4] whereas melamine-exfoliated graphene showed less than 24 hours stability in cell culture media.[54] Graphene exfoliated in BSA via ultrasonication also showed a 2-month shelf-life, but the conditions are unavailable.[50] Storage in a fridge or freezer is preferred for these materials since proteins are vulnerable to microbes

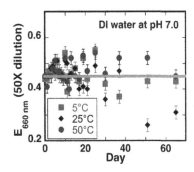

FIGURE 2.12

Storage stability of protein-exfoliated graphene.[4]

and proteases present in a nonsterile laboratory. It is advised to follow sterile conditions during the preparation of graphene for biological studies.

There is a discrepancy in finding which protein can exfoliate graphite. Lysozyme, e.g., was able to exfoliate graphite using ultrasonication and the strong cationic nature (+9 at pH 7.0) of the protein contributed to the colloidal stability of graphene. However, negatively charged ovalbumin (−15 at pH 7.0) did not show exfoliation abilities upon ultrasonication.[51] Conversely, using shear methods, ovalbumin produced graphene dispersions and lysozyme failed to do so.[4] Interestingly, BSA was successful in both methods but hemoglobin also failed. Perhaps the high-speed shearing and ultrasonication affects the protein structure differently, which influences their exfoliation efficiency. This also indicates that the kinetic trapping of exfoliated graphene (Fig. 2.9), followed by the electrostatic stabilization, is not only the function of proteins but they also facilitate the delamination process. This is possible if protein solutions affect the bulk property, such as surface tension of the medium, and they are known to do so.[62] But this aspect of protein- or surfactant-based exfoliation has not yet been explored (Table 2.7).

2.5 CONCLUSIONS

Methods for *the* preparation of graphene targeting diverse applications are known today. The synthetic routes are generally application specific, hence, a single preparative method is not enough to fit all the fields. For example, electronic devices show the best performance with defect-free, large-area graphene, which is realized by CVD. However, composite preparations require inexpensive, high-concentration graphene dispersions, preferably in water, where liquid-phase exfoliation from graphite is preferred.[9] Similarly, oxidative routes of graphite exfoliation, involving strong acids and strong oxidants, result in structurally defective graphene oxide which is not ideal for technologies that need high concentrations or large volumes. Direct, liquid-phase exfoliation is a good alternative to such needs, which can easily be scaled up to thousands of liters. However, the high polydispersity of graphene due to the fragmentation of graphene layers under mechanical forces could limit some uses of graphene dispersions.

Selection of exfoliating agents and solvents for exfoliation is another criterion required to be satisfied for specific applications. For instance, graphene exfoliated in organic solvents, such as NMP, is not useful for biological applications where aqueous conditions are needed. Likewise, surfactant-exfoliated graphene under aqueous conditions is also not amenable to applications involving living systems because of the poor biocompatibility of most of the surfactants.[69] Alternatively, exfoliation of graphite using biomolecules in water is suitable for biological uses especially proteins, such as albumins, which are usually part of the cell culture media and do not have catalytic activity. Exfoliation of graphite using bovine serum albumin is well explored using both ultrasonication and high shear-mixing

Table 2.7 Key Properties of Proteins Used for Graphite Exfoliation by Shear or Sonication

Protein	pI/net Charge at pH 7	Mol. wt. (kDa)	Charged Amino Acid Composition	Exfoliation Method
Human Serum Albumin (HSA)[63]	4.7/−15	66	Lys—59 Arg—24 Asp—36 Glu—62	Ultrasonication and shear[6]
Ovalbumin (OVL) (Chicken hen egg)[64]	4.9/−12	43	Lys—20 Arg—15 Asp—14 Glu—33	Shear[4]
α-lactoglobulin (BLG) (Bovine Milk)[65]	5.1/−18	37	Lys—30 Arg—6 Asp—20 Glu—16	Shear[4]
Bovine serum albumin (BSA)[66]	5.6/−17	66	Lys—59 Arg—23 Asp—40 Glu—59	Ultrasonication and shear[4,50]
Hydrophobin HFB1 (Class 2) (*Trichoderma reesei*)	6.7/0	7.5 (mono)	Lys—2 Arg—1 Asp—3 Glu—0	Ultrasonication[52]
Bovine Hemoglobin[67]	6.8/ + 1	62	Lys—48 Arg—14 Asp—34 Glu—26	Shear[4]

(Continued)

Table 2.7 Key Properties of Proteins Used for Graphite Exfoliation by Shear or Sonication *Continued*

Protein	pI/net Charge at pH 7	Mol. wt. (kDa)	Charged Amino Acid Composition	Exfoliation Method
 Lysozyme[68] (Chicken hen egg)	11.3/ + 8	14	Lys—6 Arg—11 Asp—7 Glu—2	Ultrasonication[51]

conditions. The highest exfoliation rates and final graphene concentration (7 mg/mL) were achieved by exfoliation of graphite in a kitchen blender using BSA, which is referred to as biographene.[4] This method offers graphene production under greener and inexpensive routes, with minimum resources as simple as a kitchen blender! Graphene production using simple methods, such as a kitchen blender or Scotch tape, have had a tremendous impact on the research and development of graphene all around the world, which is incomparable with any other existing material.

REFERENCES

1. Tour, J. M. Top-Down Versus Bottom-Up Fabrication of Graphene-Based Electronics. *Chem. Mater.* **2014,** *26*, 163—171.
2. Chen, L.; Hernandez, Y.; Feng, X.; Müllen, K. From Nanographene and Graphene Nanoribbons to Graphene Sheets: Chemical Synthesis. *Angew. Chem. Inter. Ed.* **2012,** *51*, 7640—7654.
3. Novoselov, K. S.; Geim, A. K.; Morozov, S. V.; Jiang, D.; Zhang, Y.; Dubonos, S. V.; Grigorieva, I. V.; Firsov, A. A. Electric Field Effect in Atomically Thin Carbon Films. *Science* **2004,** *306*, 666.
4. Pattammattel, A.; Kumar, C. V. Kitchen Chemistry 101: Multigram Production of High Quality Biographene in a Blender With Edible Proteins. *Adv. Funct. Mater.* **2015,** *25*, 7088—7098.
5. Kumar, C. V.; Pattammattel, A. Chapter Eleven: BioGraphene: Direct Exfoliation of Graphite in a Kitchen Blender for Enzymology Applications. In *Methods in Enzymology;* Challa Vijaya, K., Ed.; Vol. 571; Academic Press, 2016; pp 225—244.
6. Pattammattel, A.; Pande, P.; Kuttappan, D.; Basu, A.B.; Amalaradjou, A.; Kumar, C.V. Graphite exfoliation in animal sera: Scalable Production, Enhanced Stability and Reduced Toxicity. *Submitted Manuscript.*

7. Nicolosi, V.; Chhowalla, M.; Kanatzidis, M. G.; Strano, M. S.; Coleman, J. N. Liquid Exfoliation of Layered Materials. *Science* **2013,** *340,* 1226419.

8. Bepete, G.; Anglaret, E.; Ortolani, L.; Morandi, V.; Huang, K.; Pénicaud, A.; Drummond, C. Surfactant-Free Single-Layer Graphene in Water. *Nat. Chem.* **2016.** advance online publication.

9. Ferrari, A. C.; Bonaccorso, F.; Fal'ko, V.; Novoselov, K. S.; Roche, S.; Boggild, P.; Borini, S.; Koppens, F. H. L.; Palermo, V.; Pugno, N., et al. Science and Technology Roadmap for Graphene, Related Two-Dimensional Crystals, and Hybrid Systems. *Nanoscale* **2015,** *7,* 4598−4810.

10. Zhang, Y.; Zhang, L.; Zhou, C. Review of Chemical Vapor Deposition of Graphene and Related Applications. *Accounts Chem. Res.* **2013,** *46,* 2329−2339.

11. Chen, Y.; Gong, X.-L.; Gai, J.-G. Progress and Challenges in Transfer of Large-Area Graphene Films. *Adv. Sci.* **2016,** *3* 1500343-n/a.

12. Osikoya, A. O.; Parlak, O.; Murugan, N. A.; Dikio, E. D.; Moloto, H.; Uzun, L.; Turner, A. P. F.; Tiwari, A. Acetylene-Sourced CVD-Synthesised Catalytically Active Graphene for Electrochemical Biosensing. *Biosensors Bioelectron.* **2017,** *89* (Part 1), 496−504.

13. Fogel, Y.; Zhi, L.; Rouhanipour, A.; Andrienko, D.; Räder, H. J.; Müllen, K. Graphitic Nanoribbons With Dibenzo[e,l]pyrene Repeat Units: Synthesis and Self-Assembly. *Macromolecules* **2009,** *42,* 6878−6884.

14. Ciesielski, A.; Samori, P. Graphene Via Sonication Assisted Liquid-Phase Exfoliation. *Chem. Soc. Rev.* **2014,** *43,* 381−398.

15. Yi, M.; Shen, Z. A Review on Mechanical Exfoliation for the Scalable Production of Graphene. *J. Mater. Chem. A* **2015,** *3,* 11700−11715.

16. Khan, U.; O'Neill, A.; Lotya, M.; De, S.; Coleman, J. N. High-Concentration Solvent Exfoliation of Graphene. *Small* **2010,** *6,* 864−871.

17. Xu, J.; Dang, D. K.; Tran, V. T.; Liu, X.; Chung, J. S.; Hur, S. H.; Choi, W. M.; Kim, E. J.; Kohl, P. A. Liquid-Phase Exfoliation of Graphene in Organic Solvents With Addition of Naphthalene. *J. Colloid Interf. Sci.* **2014,** *418,* 37−42.

18. Hernandez, Y.; Nicolosi, V.; Lotya, M.; Blighe, F. M.; Sun, Z.; De, S.; McGovern, I. T.; Holland, B.; Byrne, M.; Gun'Ko, Y. K., et al. *High-Yield Production of Graphene by Liquid-Phase Exfoliation of Graphite. Nat. Nano,* 3. 2008563−568.

19. Lotya, M.; King, P. J.; Khan, U.; De, S.; Coleman, J. N. High-Concentration, Surfactant-Stabilized Graphene Dispersions. *ACS Nano* **2010,** *4,* 3155−3162.

20. Parviz, D.; Das, S.; Ahmed, H. S. T.; Irin, F.; Bhattacharia, S.; Green, M. J. Dispersions of Non-Covalently Functionalized Graphene with Minimal Stabilizer. *ACS Nano* **2012,** *6,* 8857−8867.

21. Zhao, W.; Fang, M.; Wu, F.; Wu, H.; Wang, L.; Chen, G. Preparation of Graphene by Exfoliation of Graphite Using Wet Ball Milling. *J. Mater. Chem.* **2010,** *20,* 5817−5819.

22. Kumar, G. R.; Jayasankar, K.; Das, S. K.; Dash, T.; Dash, A.; Jena, B. K.; Mishra, B. K. Shear-Force-Dominated Dual-Drive Planetary Ball Milling for the Scalable Production of Graphene and Its Electrocatalytic Application With Pd Nanostructures. *RSC Adv.* **2016,** *6,* 20067−20073.

23. Yan, L.; Lin, M.; Zeng, C.; Chen, Z.; Zhang, S.; Zhao, X.; Wu, A.; Wang, Y.; Dai, L.; Qu, J., et al. Electroactive and Biocompatible Hydroxyl- Functionalized Graphene by Ball Milling. *J. Mater. Chem.* **2012,** *22,* 8367−8371.

24. Liu, L.; Xiong, Z.; Hu, D.; Wu, G.; Chen, P. Production of High Quality Single- or Few-Layered Graphene by Solid Exfoliation of Graphite in the Presence of Ammonia Borane. *Chem. Communicat.* **2013**, *49*, 7890−7892.

25. Chua, C. K.; Sofer, Z.; Khezri, B.; Webster, R. D.; Pumera, M. Ball-Milled Sulfur-Doped Graphene Materials Contain Metallic Impurities Originating From Ball-Milling Apparatus: Their Influence on the Catalytic Properties. *Phys. Chem. Chem. Phys.* **2016**, *18*, 17875−17880.

26. Aparna, R.; Sivakumar, N.; Balakrishnan, A.; Nair, A. S.; Nair, S. V.; Subramanian, K. R. V. An Effective Route to Produce Few-Layer Graphene Using Combinatorial Ball Milling and Strong Aqueous Exfoliants. *J. Renew. Sustain. Energy* **2013**, *5*, 033123.

27. Posudievsky, O. Y.; Khazieieva, O. A.; Cherepanov, V. V.; Koshechko, V. G.; Pokhodenko, V. D. High Yield of Graphene by Dispersant-Free Liquid Exfoliation of Mechanochemically Delaminated Graphite. *J. Nanopart. Res.* **2013**, *15*, 2046.

28. Leon, V.; Quintana, M.; Herrero, M. A.; Fierro, J. L. G.; Hoz, A. D. L.; Prato, M.; Vazquez, E. Few-Layer Graphenes From Ball-Milling of Graphite With Melamine. *Chem. Communicat.* **2011**, *47*, 10936−10938.

29. Yu, P.; Lowe, S. E.; Simon, G. P.; Zhong, Y. L. Electrochemical Exfoliation of Graphite and Production of Functional Graphene. *Curr. Opin. Colloid Interf. Sci.* **2015**, *20*, 329−338.

30. Parvez, K.; Wu, Z.-S.; Li, R.; Liu, X.; Graf, R.; Feng, X.; Müllen, K. Exfoliation of Graphite into Graphene in Aqueous Solutions of Inorganic Salts. *J. Am. Chem. Soc.* **2014**, *136*, 6083−6091.

31. Su, C.-Y.; Lu, A.-Y.; Xu, Y.; Chen, F.-R.; Khlobystov, A. N.; Li, L.-J. High-Quality Thin Graphene Films from Fast Electrochemical Exfoliation. *ACS Nano* **2011**, *5*, 2332−2339.

32. Lu, J.; Yang, J.-x; Wang, J.; Lim, A.; Wang, S.; Loh, K. P. One-Pot Synthesis of Fluorescent Carbon Nanoribbons, Nanoparticles, and Graphene by the Exfoliation of Graphite in Ionic Liquids. *ACS Nano* **2009**, *3*, 2367−2375.

33. Zhong, Y. L.; Swager, T. M. Enhanced Electrochemical Expansion of Graphite for in Situ Electrochemical Functionalization. *J. Am. Chem. Soc.* **2012**, *134*, 17896−17899.

34. Paton, K. R.; Varrla, E.; Backes, C.; Smith, R. J.; Khan, U.; O'Neill, A.; Boland, C.; Lotya, M.; Istrate, O. M.; King, P., et al. Scalable Production of Large Quantities of Defect-Free Few-Layer Graphene by Shear Exfoliation in Liquids. *Nat. Mater.* **2014**, *13*, 624−630.

35. Varrla, E.; Paton, K. R.; Backes, C.; Harvey, A.; Smith, R. J.; McCauley, J.; Coleman, J. N. Turbulence-Assisted Shear Exfoliation of Graphene Using Household Detergent and a Kitchen Blender. *Nanoscale* **2014**, *6*, 11810−11819.

36. Yi, M.; Shen, Z. Kitchen Blender for Producing High-Quality Few-Layer Graphene. *Carbon* **2014**, *78*, 622−626.

37. Blomquist, N.; Engström, A.-C.; Hummelgård, M.; Andres, B.; Forsberg, S.; Olin, H. Large-Scale Production of Nanographite by Tube-Shear Exfoliation in Water. *PLoS One* **2016**, *11*, e0154686.

38. Narayan, R.; Kim, S. O. Surfactant Mediated Liquid Phase Exfoliation of Graphene. *Nano Converg.* **2015**, *2*, 20.

39. Haar, S.; El Gemayel, M.; Shin, Y.; Melinte, G.; Squillaci, M. A.; Ersen, O.; Casiraghi, C.; Ciesielski, A.; Samorì, P. Enhancing the Liquid-Phase Exfoliation of

Graphene in Organic Solvents upon Addition of n-Octylbenzene. *Scientific Reports* **2015,** *5,* 16684.

40. Yang, H.; Hernandez, Y.; Schlierf, A.; Felten, A.; Eckmann, A.; Johal, S.; Louette, P.; Pireaux, J. J.; Feng, X.; Mullen, K., et al. A Simple Method for Graphene Production Based on Exfoliation of Graphite in Water Using 1-Pyrenesulfonic Acid Sodium Salt. *Carbon* **2013,** *53,* 357−365.

41. Sampath, S.; Basuray, A. N.; Hartlieb, K. J.; Aytun, T.; Stupp, S. I.; Stoddart, J. F. Direct Exfoliation of Graphite to Graphene in Aqueous Media with Diazaperopyrenium Dications. *Adv. Mater.* **2013,** *25,* 2740−2745.

42. Englert, J. M.; Röhrl, J.; Schmidt, C. D.; Graupner, R.; Hundhausen, M.; Hauke, F.; Hirsch, A. Soluble Graphene: Generation of Aqueous Graphene Solutions Aided by a Perylenebisimide-Based Bolaamphiphile. *Adv. Mater.* **2009,** *21,* 4265−4269.

43. White, P. A. The Genotoxicity of Priority Polycyclic Aromatic Hydrocarbons in Complex Mixtures. *Mutat. Res./Genetic Toxicol. Environ. Mutag.* **2002,** *515,* 85−98.

44. Guardia, L.; Fernández-Merino, M. J.; Paredes, J. I.; Solís-Fernández, P.; Villar-Rodil, S.; Martínez-Alonso, A.; Tascón, J. M. D. High-Throughput Production of Pristine Graphene in an Aqueous Dispersion Assisted by Non-Ionic Surfactants. *Carbon* **2011,** *49,* 1653−1662.

45. Bourlinos, A. B.; Georgakilas, V.; Zboril, R.; Steriotis, T. A.; Stubos, A. K.; Trapalis, C. Aqueous-Phase Exfoliation of Graphite in the Presence of Polyvinylpyrrolidone for the Production of Water-Soluble Graphenes. *Solid State Commun.* **2009,** *149,* 2172−2176.

46. Paredes, J. I.; Villar-Rodil, S. Biomolecule-Assisted Exfoliation and Dispersion of Graphene and Other Two-Dimensional Materials: A Review of Recent Progress and Applications. *Nanoscale* **2016,** *8,* 15389−15413.

47. Ahadian, S.; Estili, M.; Surya, V. J.; Ramon-Azcon, J.; Liang, X.; Shiku, H.; Ramalingam, M.; Matsue, T.; Sakka, Y.; Bae, H., et al. Facile and Green Production of Aqueous Graphene Dispersions for Biomedical Applications. *Nanoscale* **2015,** *7,* 6436−6443.

48. Joseph, D.; Tyagi, N.; Ghimire, A.; Geckeler, K. E. A Direct Route Towards Preparing pH-Sensitive Graphene Nanosheets With Anti-Cancer Activity. *RSC Adv.* **2014,** *4,* 4085−4093.

49. Gravagnuolo, A. M.; Morales-Narváez, E.; Longobardi, S.; da Silva, E. T.; Giardina, P.; Merkoçi, A. In Situ Production of Biofunctionalized Few-Layer Defect-Free Microsheets of Graphene. *Adv. Funct. Mater.* **2015,** *25,* 2771−2779.

50. Ge, Y.; Wang, J.; Shi, Z.; Yin, J. Gelatin-Assisted Fabrication of Water-Dispersible Graphene and Its Inorganic Analogues. *J. Mater. Chem.* **2012,** *22,* 17619−17624.

51. León, V.; Rodriguez, A. M.; Prieto, P.; Prato, M.; Vázquez, E. Exfoliation of Graphite with Triazine Derivatives under Ball-Milling Conditions: Preparation of Few-Layer Graphene via Selective Noncovalent Interactions. *ACS Nano* **2014,** *8,* 563−571.

52. He, P.; Zhou, C.; Tian, S.; Sun, J.; Yang, S.; Ding, G.; Xie, X.; Jiang, M. Urea-Assisted Aqueous Exfoliation of Graphite for Obtaining High-Quality Graphene. *Chem. Commun.* **2015,** *51,* 4651−4654.

53. Punith Kumar, M. K.; Shanthini, S.; Srivastava, C. Electrochemical Exfoliation of Graphite for Producing Graphene Using Saccharin. *RSC Adv.* **2015,** *5,* 53865−53869.

54. Rubio, N.; Serra-Maia, R.; Kafa, H.; Mei, K.-C.; Pach, E.; Luckhurst, W.; Zloh, M.; Festy, F.; Richardson, J. P.; Naglik, J. R., et al. Production of Water-Soluble Few-

Layer Graphene Mesosheets by Dry Milling with Hydrophobic Drug. *Langmuir* **2014,** *30,* 14999−15008.

55. Joseph, D.; Seo, S.; Williams, D. R.; Geckeler, K. E. Double-Stranded DNA-Graphene Hybrid: Preparation and Anti-Proliferative Activity. *ACS Appl. Mater. Interf.* **2014,** *6,* 3347−3356.

56. Sharifi, F.; Bauld, R.; Ahmed, M. S.; Fanchini, G. Transparent and Conducting Graphene−RNA-Based Nanocomposites. *Small* **2012,** *8,* 699−706.

57. Ayán-Varela, M.; Paredes, J. I.; Guardia, L.; Villar-Rodil, S.; Munuera, J. M.; Díaz-González, M.; Fernández-Sánchez, C.; Martínez-Alonso, A.; Tascón, J. M. D. Achieving Extremely Concentrated Aqueous Dispersions of Graphene Flakes and Catalytically Efficient Graphene-Metal Nanoparticle Hybrids with Flavin Mononucleotide as a High-Performance Stabilizer. *ACS Appl. Mater. Interf.* **2015,** *7,* 10293−10307.

58. Uysal Unalan, I.; Wan, C.; Trabattoni, S.; Piergiovanni, L.; Farris, S. Polysaccharide-Assisted Rapid Exfoliation of Graphite Platelets Into High Quality Water-Dispersible Graphene Sheets. *RSC Adv.* **2015,** *5,* 26482−26490.

59. Chabot, V.; Kim, B.; Sloper, B.; Tzoganakis, C.; Yu, A. High Yield Production and Purification of Few Layer Graphene by Gum Arabic Assisted Physical Sonication. *Sci. Rep.* **2013,** *3,* 1378.

60. Carrasco, P. M.; Montes, S.; García, I.; Borghei, M.; Jiang, H.; Odriozola, I.; Cabañero, G.; Ruiz, V. High-Concentration Aqueous Dispersions of Graphene Produced by Exfoliation of Graphite Using Cellulose Nanocrystals. *Carbon* **2014,** *70,* 157−163.

61. Liu, W.; Zhou, R.; Zhou, D.; Ding, G.; Soah, J. M.; Yue, C. Y.; Lu, X. Lignin-Assisted Direct Exfoliation of Graphite to Graphene in Aqueous Media and Its Application in Polymer Composites. *Carbon* **2015,** *83,* 188−197.

62. Kitabatake, N.; Doi, E. Surface Tension and Foaming of Protein Solutions. *J. Food Sci.* **1982,** *47,* 1218−1221.

63. Wang, Z.-m; Ho, J. X.; Ruble, J. R.; Rose, J.; Rüker, F.; Ellenburg, M.; Murphy, R.; Click, J.; Soistman, E.; Wilkerson, L., et al. Structural Studies of Several Clinically Important Oncology Drugs in Complex With Human Serum Albumin. *Biochim. Biophys. Acta (BBA) - General Sub.* **2013,** *1830,* 5356−5374.

64. Stein, P. E.; Leslie, A. G. W.; Finch, J. T.; Carrell, R. W. Crystal Structure of Uncleaved Ovalbumin at 1·95 Å Resolution. *J. Mol. Biol.* **1991,** *221,* 941−959.

65. Brownlow, S.; Cabral, J. H. M.; Cooper, R.; Flower, D. R.; Yewdall, S. J.; Polikarpov, I.; North, A. C. T.; Sawyer, L. Bovine β-Lactoglobulin at 1.8 Å Resolution — Still an Enigmatic Lipocalin. *Structure* **1997,** *5,* 481−495.

66. Majorek, K. A.; Porebski, P. J.; Dayal, A.; Zimmerman, M. D.; Jablonska, K.; Stewart, A. J.; Chruszcz, M.; Minor, W. Structural and Immunologic Characterization of Bovine, Horse, and Rabbit Serum Albumins. *Mol. Immunol.* **2012,** *52,* 174−182.

67. Aranda, R.; Cai, H.; Worley, C. E.; Levin, E. J.; Li, R.; Olson, J. S.; Phillips, G. N.; Richards, M. P. Structural Analysis of Fish Versus Mammalian Hemoglobins: Effect of the Heme Pocket Environment on Autooxidation and Hemin Loss. *Prot. Struct. Funct. Bioinform.* **2009,** *75,* 217−230.

68. Diamond, R. Real-Space Refinement of the Structure of Hen Egg-White Lysozyme. *J. Mol. Biol.* **1974,** *82,* 371−391.

69. Blasco, J.; Hampel, M.; Moreno-Garrido, I. *Chapter 7 Toxicity of Surfactants. Comprehensive Analytical Chemistry,* Volume 40. Elsevier, 2003827−925.

Characterization techniques for graphene

3

CHAPTER OUTLINE

3.1 Summary 45
3.2 Introduction 46
3.3 Spectroscopic Methods 46
    3.3.1 Absorption/Extinction Spectroscopy 47
    3.3.2 Raman Spectroscopy 50
    3.3.3 IR Spectroscopy 55
    3.3.4 X-Ray Photoelectron Spectroscopy 56
    3.3.5 X-Ray Diffraction (XRD) 57
    3.3.6 Zeta Potential Measurements 59
    3.3.7 Surface Plasmon Resonance Spectroscopy 61
3.4 Microscopy Methods 63
    3.4.1 SEM and TEM 63
    3.4.2 Atomic Force Microscopy/Scanning Tunneling Microscopy 65
    3.4.3 Fluorescence Quenching Microscopy 67
3.5 Conclusions 67
References 69

3.1 SUMMARY

Techniques for the thorough characterization of graphene are discussed in this chapter. The main spectroscopic and microscopic methods are examined in depth. The center of the discussion is how to distinguish graphene, graphene oxide, and reduced graphene oxide, which are frequently (inappropriately) termed "graphene." The second goal of these characterizations is to analyze the size, number of layers, and extent of defects. The spectroscopic methods are reviewed first, including absorption, Raman, X-ray photoelectron, and infrared (IR) spectroscopy. Among these, Raman spectroscopy stands out, as it is recognized as the "fingerprint technique" to fully characterize graphene under ambient conditions. The capabilities of these methods to examine graphene suspended in a liquid or directly in the solid state were specifically considered, which is crucial for biological applications of graphene. Additionally, the chances of errors with these methods are considered, especially in quantitative analysis, due to drying, aggregation, and interference from exfoliating or stabilizing agents. Some suggestions to

Introduction to Graphene. DOI: http://dx.doi.org/10.1016/B978-0-12-813182-4.00003-9

minimize these errors are also included. Subsequently, the microscopic techniques for direct visualization of graphene sheets are reviewed. Although these techniques cannot directly distinguish graphene from its chemically oxidized counterparts, the thickness, lateral size, basal plane modifications, but the elemental compositions can be obtained when the microscopy is coupled with energy-dispersive X-ray spectroscopy. Powerful techniques, such as scanning tunneling microscopes, can scan graphene layers to visualize hexagonal rings in the plane. Combining all these techniques, the nature of graphene and its composites are more fully analyzed at the atomic level, which enables designing and predicting novel engineered graphene materials for specific applications.

3.2 INTRODUCTION

Graphene layers that are produced by chemical vapor deposition (CVD) or by exfoliation methods need to be characterized using multiple spectroscopic and microscopic techniques prior to their use in any application or for fundamental studies.[1] There are several benchmark techniques to determine the success of exfoliation or graphene growth on surfaces, the number of layers present, size (area, width, length) and the extent as well as the type of defects present in the sheets.[2] These methods provide convenient handles to distinguish graphene from its counterparts such as graphene oxide and reduced graphene oxide. Spectroscopic and microscopic data are combined to examine the above characteristics of graphene as well as surface functionalization of the sheets. Spectroscopic techniques offer characterization of graphene in the bulk form in solution, as powders as well as in composite forms.

3.3 SPECTROSCOPIC METHODS

The spectroscopy techniques that have been developed to characterize graphene include absorption/extinction spectroscopy,[3] Raman spectroscopy,[4] X-ray photoelectron spectroscopy (XPS),[5] Fourier transform infrared (FTIR),[6] fluorescence spectroscopy,[7] and so on. Empirical equations are developed for the interpretation of the absorption and Raman spectroscopy data and to determine the microstructure of graphene.[2,3] These techniques offer in-line, high-throughput methods to characterize graphene, especially when produced using liquid-phase exfoliation methods. XPS and FTIR were used to find the ratio of the number of sp^2 to sp^3 carbons. XPS can identify carbons bonded to oxygen or other carbons, whereas FTIR spectra categorize functional groups present on graphene sheets. These techniques are important to analyze the extent of oxidation of graphene during the synthesis or due to its exposure to solvents or ambient conditions. Fluorescence microscopy was used to calculate the surface area of graphene derivatives based on dye adsorption isotherms. This indirect method helps to find the surface area of graphene in suspension, whereas isotherms based on BET (Brunauer–Emmett–Teller) work with solid samples. The field of graphene characterization has been expanding so rapidly that simple and commonly available absorption spectrophotometers can quantify microfeatures of graphene, and such methods promise the mass production of graphene and its quality assessment.

3.3.1 ABSORPTION/EXTINCTION SPECTROSCOPY

Graphene sheets interact with radiation of 200–800 nm by both absorption and scattering. The extinction spectroscopy is used extensively to characterize graphene and its dispersions in the liquid phase.[3] The major absorption peak is around 260–270 nm which is due to $\pi-\pi^*$ transitions of the conjugated aromatic rings and extend well into the near-infrared region (scattering + absorbance) (Fig. 3.1). If a protein or a biomolecule is used to exfoliate graphene, then the contributions of the biomolecule to the absorbance, in this range must be considered. The absorbance of graphene prepared with bovine serum albumin as the exfoliating agent under shear conditions is shown in Fig. 3.1A.[8] The absorbance of the sample scaled linearly with concentration (Fig. 3.1B). The absorption and scattering of light by the sheets at longer wavelengths depend on the thickness of the sheets.[3] Chemically oxidized graphene oxide shows a much broader peak in the UV region due to the overlap of the $\pi - \pi^*$ and $n - \pi^*$ transitions.[9] The peak of graphene (oxide) at around 600 nm is primarily due to scattering and it is used to quantitate concentrations in suspensions (Eq. 3.1). The extinction of coefficient of particular graphene suspensions are often calculated by measuring the absorbance of a series of known concentrations of graphene suspensions (mg/mL). The

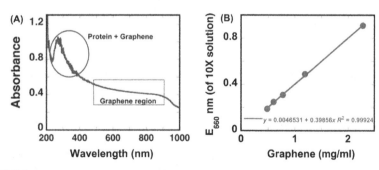

FIGURE 3.1

(A) Absorption spectrum of graphene exfoliated and stabilized by bovine serum albumin (B) Extinction of graphene suspension as a function of concentration (sample was exfoliated in BSA (3 mg/mL) in water at pH 7.4 using a kitchen blender) at 660 nm, follows Beer Lambert's law.[8] © John Wiley & Sons.

observed absorbance in this range can vary from sample to sample due to the different contributions of concentration as well as scattering due to the differences in the thicknesses and sizes of the sheets as well. Thus, there is no universal extinction coefficient for graphene extinction at this range, and the coefficient is dependent on the exfoliating agent and the method of exfoliation.

$$E = \alpha \cdot c \cdot l \tag{3.1}$$

E, absorption and scattering (extinction); α, extinction coefficient; c, concentration; l, path length.

The extinction coefficient of aqueous graphene suspensions is highly variable, varying with the kind of exfoliating agents used. Table 3.1 examines the values reported, when exfoliation was carried out in water. The major reason for this variability is due to the variations in the thickness (number of sheets stacked together) and their average size, as well as the adsorption of exfoliating agent on the sheets (Fig. 3.2A). The role of solvent in affecting the extinction of graphene sheets (hypo- or hyperchromic shifts) is also possible. The thickness and size intrinsically enhance extinction, whereas binding of the exfoliating agent increases the effective mass of graphene.[3] One way to correct for the latter is by quantitating the amount of the bound species either by assays or by elemental analysis. Qualitative analysis of graphene by the extinction spectra has been developed recently.

The ratio of graphene $\pi - \pi^*$ absorption band (260–280 nm) and extinction in the plateau region (550 nm) was used to calculate the number of layers (thickness) of graphene sheets within $\pm 15\%$ accuracy. The graphene sheets were exfoliated in sodium cholate by the shear method, and thickness was measured using

Table 3.1 Measured Extinction Coefficient (660 nm) and Zeta Potential of Graphene in Water

Exfoliation Method	Exfoliating Agent	Extinction Coefficient, α (mL/mg/m)	Zeta Potential (mV)
Shear/turbulence	Sodium cholate	6600[10]	NA
Ultrasonication	Gum arabic	5422[11]	NA
Ultrasonication	Ionic liquid copolymer	4890[12] (500 nm)	NA
Ultrasonication	Chitosan	2287[14]	+43 (pH 3.5)
Ultrasonication	Sodium dodecylbenzene sulfonate (SDBS)	1390[13]	−44 (pH 7.4)
Ultrasonication	Pullulan	1240[14]	+2.2 (pH 5)
Ultrasonication	Alginate	525[14]	−82 (pH 5)
Shear/Turbulence	Bovine serum albumin or animal sera	398[8]	−25 (pH 7.4)

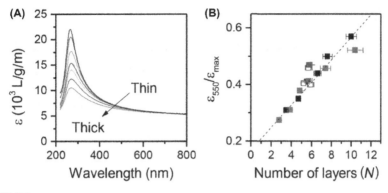

FIGURE 3.2

(A) Dependence of extinction and thickness of size-separated graphene (cited as graphene oxide above) on dispersions in sodium cholate. (B) Linear relationship between extinction of the $\pi - \pi^*$ transition and plateau region at 550 nm.

Reproduced from Backes, C., Paton, K.R., Hanlon, D., Yuan, S., Katsnelson, M.I., Houston, J., et al. Spectroscopic metrics allow in situ measurement of mean size and thickness of liquid-exfoliated few-layer graphene nanosheets. Nanoscale. 2016;8(7):4311–4323 with permission. © Royal Chemical Society.

atomic force microscopy (AFM).[3] The correlation of thickness and extinction put forward the following empirical equation:

$$<N> = (25 \times E_{550}/E_{\pi-\pi^*}) - 4.2 \tag{3.2}$$

where $<N>$ is the average number of layers for $2.7 \leq <N> \leq 10.4$, and E is the extinction at specific wavelength.

However, the absorbance measurement at the $\pi - \pi^*$ transition of graphene can have severe interference due to exfoliating agents with aromatic or carbonyl groups $(n - \pi^*)$. When the absorbance at 325 nm was used instead of 270−280 nm, the following equation was fitted within $\pm 20\%$ accuracy:[3]

$$<N> = (35.7 \times E_{550}/E_{325}) - 14.8 \quad 2.7 \leq <N> \leq 10.4. \tag{3.3}$$

Such equations are very useful in characterizing graphene for mass production. But the parameters are very sensitive to the microenvironment of graphene, as well as the thickness of the flakes.[3] Therefore adjustments may be required for specific exfoliating agents, solvents, or methods used for exfoliation. For an illustration, Eq. 3.3 was used to calculate the thickness for graphene produced in animal sera by shear exfoliation (Table 3.2).[15] The sample indicated 3−5-layer graphene by Raman spectral analysis. After fitting the absorption data at 550/325 nm to Eq. 3.3, we noticed that some sets resulted in negative numbers (samples were exfoliated in animal sera (10% by volume) in water at pH 7.4 using a kitchen blender) indicating errors due to the solvent effect or due to the protein, although sera do not absorb significantly at these wavelengths. Nevertheless, this approach may be modified according to the microenvironments

Table 3.2 Calculated Number of Layers of Graphene Exfoliated in Animal Sera,[15] Using Eq. 3.3

Serum Type	Kitchen Blender	Shear	Sonication
Bovine	3.4 ± 0.5	2.2 ± 0.7	2.2 ± 0.3
Chicken	4.8 ± 0.4	0.9 ± 0.6	1.4 ± 0.4
Horse	3.1 ± 0.4	2.2 ± 0.7	0.5 ± 0.5
Human	1.7 ± 0.4	*−0.6 ± 0.3*	1.3 ± 0.5
Porcine	3.9 ± 0.5	1.01 ± 0.4	*−1.1 ± 0.4*
Rabbit	1.5 ± 0.9	6.8 ± 0.4	3.4 ± 1.0

Negative numbers indicate errors in measurements even after averaging 10 measurements (shown in italics). The equation needs to be used with caution.

to generate a standard curve to enable fast and easy characterization of graphene suspensions produced under specific conditions. Combining these data with Raman spectroscopy, another spectroscopic technique to calculate number of layers, size, and defects, meaningful information can be accessed in a short time frame.

3.3.2 RAMAN SPECTROSCOPY

Raman spectroscopy is a versatile technique to characterize graphene, comprehensively, both in bulk or in microfocus mode.[16] This nondestructive technique is suitable to both laboratory- and mass-produced samples, which brings out the atomic-scale features of graphene. Raman active bands in graphite are sensitive to the number of layers in the flake, size and degree of defects, type of defects, functionalization, doping, orientation, etc.[4] In the absence of a band gap in graphene, all the incident wavelengths resonate, which has information about both atomic and electronic levels.[17] The peak positions, shape, and intensity of Raman bands of exfoliated graphene vary with disorder, oxidation, doping, number of layers, flake size, and microenvironment around the flake. Raman is used as a quantitative measure to calculate these features and it is often in agreement with microscopy data.[18]

Raman spectra of graphene has three main peaks and several minor peaks, which contain information about structural and electronic properties. The most prominent Raman active band in graphene (Fig. 3.3A) and other sp^2 carbon allotropes is the G band ($\sim 1580 \text{ cm}^{-1}$), which arises due to the C−C in-plane vibrations (stretching mode). The D band ($\sim 1350 \text{ cm}^{-1}$) is due to the disordered carbons, arises from the breathing vibration of six-atom rings. This mode is activated only by defects or disorders, and thus, the ratio of the D to G band intensities (I_D/I_G) is used as a proxy to represent the level of defects present in the graphene preparation.[20] 2D band at $\sim 2700 \text{ cm}^{-1}$, does not require defect activation and thus is present in graphene and graphite as well. Both the D and 2D band positions depend on the excitation wavelength (dispersive bands), whereas

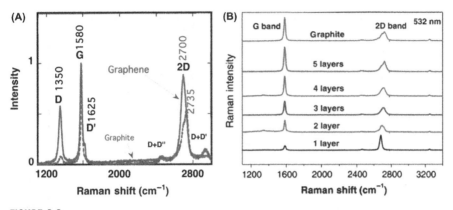

FIGURE 3.3

(A) Raman spectra of graphene exfoliated using a protein[8] (*solid line*) and graphite (*dotted line*). © John Wiley & Sons. (B) Raman spectra as a function of number of layers[19] © Springer.

the G band is nondispersive. As the number of stacks in graphene increases to few-layer graphene, the 2D band gets broader, shorter, and shifts to lower wavenumbers due to changes in the electronic environment resulting from interactions between adjacent layers.[16] As a result, after five layers, the Raman spectra of graphene and graphite are not distinguishable (Fig. 3.3B).

The Raman data were interpreted by perturbation theory where electron–photon and electron–phonon matrices at the k-space were used for undoped, unstrained, defect-free monolayer graphene.[21] The lattice of graphene is the honeycomb with hexagonal sp^2 hybridized carbon atoms. The points in reciprocal lattice of graphene, which is also the hexagonal Brillouin zone, are denoted as K or K'. The space lattice in reciprocal space is called k space, which is dispersed over the hexagonal cell in graphene.[22] Graphene has cone-like Dirac points in momentum space with K or K' being the transition point of (Fermi level) the dispersion points of valence and conduction band.[23] Low-intensity peaks in graphene are either overtones or combination tones of major peaks usually distinguishable, present in high-quality samples (Fig. 3.3A). For example, D′ peak, appearing as a shoulder of the G band, is a double resonance peak produced by the intervalley scattering from defects.[4] Thus, this peak appears only if there are defects in graphene. The ratio of intensity of D′ to D can be used to identify the edge defects, vacancies, or sp^3 defects in graphene samples (Fig. 3.4B).

The matrices of Raman intensity and microscopy data were used to calculate properties of graphene to a good approximation. Upon exfoliation of graphite, the defect peak intensity increases as more edges and vacancies are formed during exfoliation. Thus, this change was used to follow exfoliation and the quality of graphene. The I_D/I_G ratio of highly defective graphene oxide is usually above 1, while those of the exfoliated graphene derivatives range from 0.1 to 0.6. Table 3.3 lists the quality of graphene exfoliated under aqueous conditions.

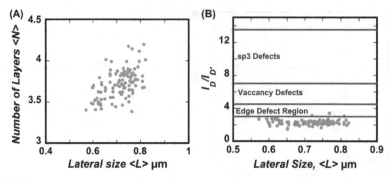

FIGURE 3.4

(A) Estimated lateral size and number of layers of graphene exfoliated with bovine serum in a kitchen blender. (B) Nearly defect-free nature of graphene is shown by $I_D/I_{D'}$ values. (sample was exfoliated in bovine serum (10%) in water at pH 7.4 using a kitchen blender).

Table 3.3 I_D/I_G Ratio of Graphene Measured in Raman Spectra Is Used to Illustrate the Quality of Graphene

Exfoliation Method	Exfoliating Agent	I_D/I_G Ratio	Excitation Wavelength (nm)
Shear/turbulence[10]	Sodium cholate (SC)	0.4 (estimate from spectrum)	N/a
Ultrasonication[11]	Gum arabic	0.3 (estimate from spectrum)	N/a
Ultrasonication[13]	Ionic liquid copolymer		
Ultrasonication[13]	Sodium dodecylbenzene sulfonate (SDBS)	0.4 (estimate from spectrum)	N/a
Ultrasonication[14]	Pullulan	0.33	514.5
Sonication[24]	Pyrene surfactant	0.33	633
Eutectic system (glove box) intercalation[25]	Ternary KCl-NaCl-ZnCl$_2$	0.15	514
Grinding in ionic liquid[26]		0.23	514
Sonication of graphite[27]	Surfactant: NMP and o-DCB	1.5	514
Mild sonication[11]	Gum arabic	0.25	633
Sonication[28]	Nonionic polymeric surfactant	0.35	N/a

(Continued)

Table 3.3 I_D/I_G Ratio of Graphene Measured in Raman Spectra Is Used to Illustrate the Quality of Graphene *Continued*

Exfoliation Method	Exfoliating Agent	I_D/I_G Ratio	Excitation Wavelength (nm)
Dissolution in superacids[29]	Chlorosulfonic acids	0.1–0.5	514
Sonication in SDBS surfactant[27]	SDBS	0.4	532
Grinding and mild sonication in ionic liquid[30]	HMIH (ionic liquid)	0.18	532
Sonication[31]	SC	0.57	633
Sonication[32]	SC	0	488
Microwave intercalation[33]	$FeCl_3$ and CH_3NO_2 cointercalated graphite	0	N/a
Sonication[34]	Nonionic block copolymers	0.45	514
Mild sonication[35]	Intercalation with ICl and IBr	0.2	633
Sonication in water–acetone mixture[36]		0.25	514
Solvothermal reduction of chemically exfoliated graphene sheets[37]		1	633
Mild sonication[38]	Perylene bisimide	0.4–0.6	532
Thermal treatment of graphite–liquid crystal composite[39]		0	633
Electrochemical exfoliation[40]	Oxone	1.24	N/a
Shear/turbulance[8]	Bovine serum albumin or animal sera	0.6	514
Ultrasonication[41]	Alginate	0.18	514.5
Ultrasonication[41]	Chitosan	0.49	514.5
Sonication[42]	MCT (myrtan condensed tannins)	0.27	532
Dissolution of potassium graphite in THF[43]	NA	1.5 (estimate from spectrum, measured in water)	532

All samples were produced or dispersed in water.

The lateral size of graphene sheets is correlated with the intensity of the G band (I_G) and D band (I_D), which needs to be normalized with the I_D/I_G value of graphite. The empirical formula for this measure is:[2]

$$\text{Lateral Size (nm)}, <L> = \frac{k}{(I_D/I_G)_{\text{Graphene}} - (I_D/I_G)_{\text{Graphite}}} \tag{3.4}$$

where $k = 0.17$, measured experimentally.[2]

Similarly, using the intensity ratio of 2D peak of graphene (~ 2700 cm^{-1}) and its shoulder (~ 2730 cm^{-1}), the number of layers of exfoliated graphene is estimated with respect to graphite. Exfoliation of graphene results in change in peak position of the 2D band to lower wavenumbers by about 30 cm^{-1}, when compared to that of graphite. For example, if the 2D peak of graphite is at 2730 cm^{-1}, the shoulder peak appears at 2700 cm^{-1}. A symmetrical 2D peak of graphene, unlike an asymmetrical graphite 2D peak, is also noted after exfoliation (Fig. 3.3A). These changes are dependent on the number of layers in the flakes, and hence, this spectroscopic marker is used to quantitate the number of layers using Eq. 3.5:[2]

$$\text{Number of layers}, <N_G> = 10^{0.84M + 0.45M^2}; M = \frac{(I_{2D}/I_{2Ds})_{\text{Graphene}}}{(I_{2D}/I_{2Ds})_{\text{Graphite}}} \tag{3.5}$$

where I_{2D} is the intensity of the 2D peak and I_{2Ds} is the intensity of the 2D shoulder peak in graphene or graphite (ω, 30 cm^{-1}).

Eq. 3.5 was originally developed using graphene that has been exfoliated with sodium cholate under high shear conditions. Interestingly, this equation was useful to analyze graphene exfoliated with hydrophobins,[44] bovine serum albumin,[8] and animal sera (Fig. 3.4A).[15] Thus, the Raman data are independent of the exfoliation medium, unlike the extinction-based equation (Eq. 3.3) discussed earlier which resulted in negative numbers in some specific instances.

Experimental considerations: Raman spectroscopy is a powerful technique, operates under ambient conditions, and samples can be in solution or deposited solid samples. A few experimental considerations while preparing samples, especially for exfoliation under conditions that are suitable for biological applications, are listed here.[45,46]

1. Large-area films of deposited graphene suspensions and collecting about 100 spectra at different locations will represent the average properties of the graphene sample.
2. Soft materials, such as proteins, used for exfoliation can be damaged by the laser beam, and so, removal of the unbound protein before film preparation improves the measurements.
3. Reducing the LASER intensity, or focusing out the light beam, reduces the damage to the sample, and could provide a better spectrum. In our experience, 50% reduction in intensity and 50% off focus of the beam (Reinshaw microRaman 2000: 514 nm excitation, 1 μm beam size and 5 mW power)

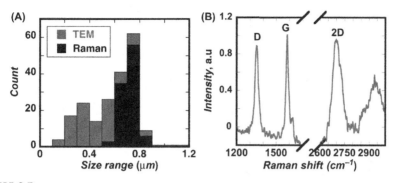

FIGURE 3.5

(A) Differences in the mean size of graphene flakes measured by Raman spectroscopy and TEM images. (B) Raman spectra of smaller graphene sheets (> 200 nm) show differences from the spectra recorded with the bulk samples (Fig. 3.3A).

gave the best spectra for graphene prepared with bovine serum albumin and other proteins.

Although Raman spectra of graphene contain a large amount of information, few limitations or opportunities for errors are worth mentioning for practical applications.

1. Most of the Raman characterization of exfoliated samples is done after depositing the layers onto a glass slide, on which restacking of the sheets can occur, which results in overestimating the number of layers. Collecting Raman spectra of films prepared from different stock solutions under dilute conditions can reduce this problem, but compromises the signal-to-noise ratio.
2. During the spectral acquisition, some samples are prone to photoreduction, especially graphene oxide samples, which can lead to the wrong conclusions. As we mentioned earlier, reducing laser power or defocusing decreases sample damage but inspection of the sample after measurements under a light microscope must be made to assess damage to the sample.
3. The empirical equations described above are useful to obtain the size information, but they cannot account for small particles (<200 nm) formed during exfoliation (Fig. 3.5). Microscopy evaluations and absorption spectroscopy measurements must be made to assess the presence of these smaller flakes.

3.3.3 IR SPECTROSCOPY

IR is often used to identify the functional groups in molecules and it is a powerful method to evaluate the extent of functionalization of graphene, graphene oxide, and chemically modified graphene.[47] In comparison to Raman spectroscopy, IR spectra of graphene derivatives do not reveal electronic or atomic features of

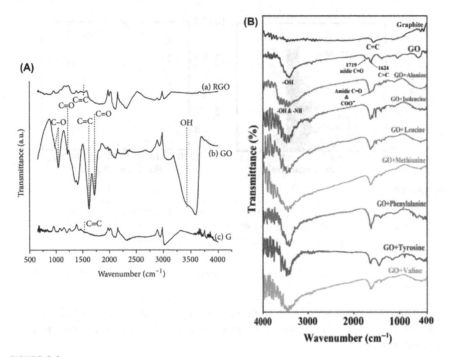

FIGURE 3.6

(A) IR spectrum of graphene oxide (GO), reduced GO (RGO) in comparison with graphite (G) (© Hindawi Publishers).[48] (B) IR spectra of amino acid functionalized graphene oxide (© Elsevier).[53]

graphene. IR spectra of graphene oxide, rich in oxygenated functional groups, identify the presence of $-OH$, $-COOH$, $C=O$, and $C-O$ functional groups in samples (Fig. 3.6A).[48] Surface functionalization including reduction,[48] reaction with amines,[49] polymers,[50] phosphates,[51] carboxylic acids,[52] and amino acids (Fig. 3.6B)[53] is characterized by IR spectroscopy. This is done by classifying the specific vibrations of functional groups and their fingerprint regions. It may be difficult to distinguish between graphite and graphene with only trace amounts of functional groups by IR spectroscopy. However, using high-energy X-rays, minor defects or oxidation are detected by XPS.

3.3.4 X-RAY PHOTOELECTRON SPECTROSCOPY

X-ray photoelectron spectra contain information about ionization energies of elements and, hence, it is useful to determine the elemental composition, doping, defects, and oxidation of graphene with high accuracy.[5,54] Binding energy of C1s electrons are usually detected around 285 eV, which is variable with the chemical state of carbon. Binding energies of common chemical states of carbon

Table 3.4 Binding Energies of C1s in Graphene Derivatives[55–58]

Chemical State	~Binding Energy, C1s (eV)
C–C	284.8
C=C	283.4
C–O	285.5
O–C–O	286.5
C=O	289
O=C–O	288.5
C–H	285.3
C–N	283.7 (sp^3 N)
C=N	286.8
$\pi - \pi^*$ resonance	291.2

compounds with oxygen (Table 3.4) are used to distinguish graphene from graphite, graphene oxide, or reduced graphene oxide (Fig. 3.7). Apart from the identification of carbon and oxygen bonding in graphene, heteroatom-doped graphene derivatives also show specific regions in XPS, which is not achievable by other techniques. The area under the curve in XPS peaks is used to quantitate the particular group which is useful to study disorders, functionalization, doping, or surface impurities in the graphene sample.

While XPS is a sensitive surface characterization technique, contamination of the sample by hydrocarbons is the biggest challenge. Atmospheric hydrocarbons from solvents, pump oils, and vacuum greases containing long-chain hydrocarbons are common contaminants, especially because of their affinity towards graphene. Oxygen-containing functional groups are sensitive to X-ray beams, and they are prone to reduction during the measurement which must also be addressed. When compared to Raman spectroscopy or extinction spectroscopy, which require only ambient conditions, XPS requires a high vacuum state, and thus the characterization is possible only in the dry state, which may not represent the bulk sample. The tedious and expensive operation of XPS instrumentation hampers collecting large numbers of spectra or analysis of large sets of data when compared to other spectroscopic techniques which are more routinely available.

3.3.5 X-RAY DIFFRACTION (XRD)

The layered packing of the honeycomb lattice of sp^2 carbons in graphite shows an interlamellar separation of 3.4 Å (001 plane),[60] demonstrated by the XRD pattern of graphite. Upon exfoliation, oxidation, or intercalation of reagents in the galleries, the interlayer distance increases, which can be characterized in bulk solids, under ambient conditions. For example, oxidation of graphite by Hummers' process resulted in interlayer spacing of ~7.4 Å (Fig. 3.8A),[61] as a result of oxygenated functional groups and trapped water molecules.[62] Intercalation of atoms and

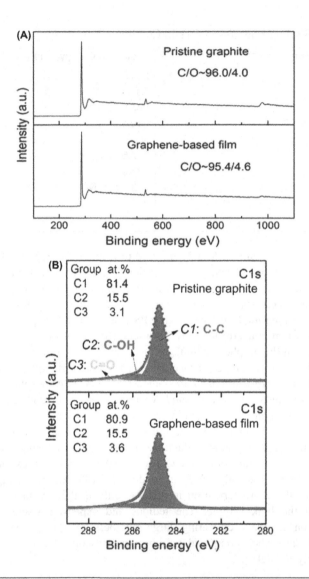

FIGURE 3.7

(A) XPS spectra of graphene exfoliated in organic solvent using a kitchen blender in comparison to that of graphite, which shows the absence of significant oxidation of the sample during exfoliation. (B) Identification of percentage composition of carbon bonding in the same sample C1 is C—C, C2 is C—O, and C3 is C═O carbon species.[59]

small molecules into graphite sheets also results in the expansion of interlayer structures. For example, phosphoric acid intercalation (nonoxidative) at 120°C (24 hours) resulted in a new phase with a layer spacing of $\sim$7.3 Å, which was further exfoliated to graphene in dimethyl formamide by ultrasonication.[63]

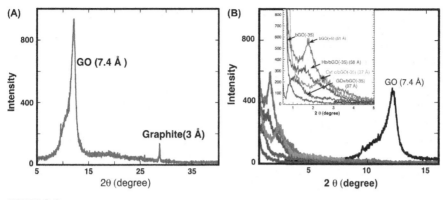

FIGURE 3.8

(A) XRD pattern of graphene oxide with a peak at 7.4 Å and the peak corresponding to graphite at higher angles. (B) XRD patterns of the protein-bound GO sheets which showed d-spacings that correlated with the corresponding protein diameters (*Hb*, hemoglobin; *GOx*, glucose oxidase; *Cyt* c, cytochrome *c*.; *bGO*, biophilized GO)[61] used for the exfoliation of graphene. © American Chemical Society.

Sample preparation for XRD requires careful consideration of the large area sampling and film preparation. Suspensions of samples are dried on glass substrates to give films of appropriate thickness and the diffraction patterns are recorded. We showed that macromolecules such as proteins can be intercalated into GO layers and these samples were examined by powder XRD. The protein/graphene suspensions coated on a glass slide are allowed to dry at room temperature, over 24 hours (Fig. 3.8B).[61] However, drying of exfoliated graphene suspensions may cause restacking of the layers, leading to phase separation of the exfoliation agent. Also, graphene exfoliated in high boiling solvents such as *N*-methyl pyrrolidone (NMP) would take longer drying periods to get a substantial amount in the film state. However, such samples may be more suitable for evaluation by alternative solution phase techniques.

3.3.6 ZETA POTENTIAL MEASUREMENTS

Particles suspended in liquids are stabilized against aggregation followed by precipitation by significant amount of charge on their surfaces. The particle surface charge attracts counter ions in the liquid and forms an ionic atmosphere around the particle. The potential between the particle surface and the liquid surrounding it is referred to as the zeta potential. When this potential is near zero, the particle has essentially no charge and is likely to aggregate and precipitate out of the solution.[64] Thus, zeta potential measurements are important in understanding the colloidal stabilities of particles suspended in liquids. The zeta potential is measured by examining the mobility of the particles subjected to an external electric field (electro-kinetic phenomena). Thus, the electro-kinetic phenomena of charged surfaces in liquids are accessed by charge—discharge cycles in colloidal suspensions.[65] Zeta potential measurements were used to determine the colloidal stability of graphene suspensions in water.

The stability of graphene suspensions in liquids is vital for applications or use in the laboratory. The stabilities of the graphene suspension stability and presence of charge on the layers are characterized using zeta potential measurements. Colloidal stability of graphene in suspensions can be evaluated by measuring the surface charge on graphene at specific pH. In general, zeta potentials above +20 mV or below −20 mV are considered as moderately stable colloidal dispersions,[66] because of significant charge repulsions between the layers. Intrinsically, unoxidized, pristine graphene should not possess any surface charge, but association of solvent or exfoliating agent on its surface gives rise to substantial zeta potential. Graphene oxide with ionizable functional groups show zeta potential of around −50 mV (at pH 7.0), and the suspensions are highly stable.[67] Exfoliated graphene suspensions showed zeta potential which depends on the type of exfoliating agent used for the graphene preparation and stabilization. When graphite was exfoliated with bovine serum albumin in a shear reactor, zeta potential of ~ -25 mV was observed at pH 7.0 and the potential was tunable by pH (Fig. 3.9A).[68] The colloidal stability of exfoliated graphene was examined by zeta potential measurements, coupled with extinction spectroscopy.

High stability of graphene dispersions in water was achieved by adding suitable exfoliating agents, which stabilize the flakes by solubilizing them in the solvent. Few-layer graphene dispersions in water had stabilities up to 2 months,[8] and this was proved using zeta potential measurements under different pH conditions.

The zeta potential of graphene under all pH conditions was unchanged after 1 week (Fig. 3.9B). However, at pH ~ 5, where the overall charge was zero, precipitation of some graphene after a week was noted (Fig. 3.9C). Thus, this characterization method measures the surface charge of graphene in suspension, therefore, graphene prepared under dry conditions or poorly dispersible in water

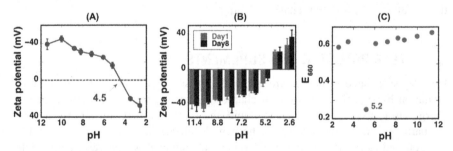

FIGURE 3.9

Change in zeta potential with pH (A) and time (B) of graphene exfoliated with bovine serum albumin in a kitchen blender.[8] (C) Retention of graphene extinction in suspension (E_{660}) over 1 week under different pH conditions. © John Wiley & Sons.

is not amenable to this measurement. To evaluate the pH stability of graphene, zeta potential is useful to study the pH dependence of the charge on graphene and for the determination of the isoelectric point (pI), pH at which the net charge is zero. Since zeta potential experiments use laser excitation, the potential for the reduction of graphene oxide during the measurement is a concern.

3.3.7 SURFACE PLASMON RESONANCE SPECTROSCOPY

The interaction of photons with the surface electrons in a metal lattice can set them into resonance, under specific conditions. When in resonance, energy can be exchanged between the radiation and the surface electrons (plasmons), which is referred to as the surface plasmon resonance (SPR). One of the resonance conditions is achieved when the momentum vectors of the plasmon and photon, along the metal surface, are exactly equal.[69] Thus, the resonance condition depends on the angle of incidence, wavelength of light, and refractive index of the medium. At fixed wavelength and refractive index, resonance condition is achieved at a specific angle, called the SPR angle (θ_{SPR}). Under these conditions, part of the incident radiation is absorbed by the metal surface and the intensity of the reflected light is reduced. The minimum reflectivity is then observed at the SPR angle. Since the resonance condition also depends on the refractive index of the medium adjacent to the metal surface, any ligand binding to this surface will alter the resonance condition, and hence change the intensity of the reflected light. Therefore, the SPR signal is extremely sensitive to the change in dielectric medium[69] and the thickness of the adsorbed layer.[70] Based on this concept, the SPR sensor chips were coated with single-layer or multilayer graphene and graphene characterized by SPR for the first time.[71]

Single-layer graphene (from ACS materials) was deposited onto the gold surface of the commercial SPR sensor chips by well-established transfer methods.[72] The thickness and morphology of single-layered graphene (SLG) on the Au surface has been examined by AFM (Fig. 3.10A) and by Raman spectroscopy (Fig. 3.10C) methods. In the Raman spectrum, the high intensity of 2D peak ($\sim 2700 \, cm^{-1}$) with respect to the G band ($\sim 1580 \, cm^{-1}$) confirms the presence of SLG.[16] The thickness of SLG on the Au surface was estimated using both AFM height trace (1.5−1.7 nm) and compared that with from the SPR method. This overestimation of SLG thickness using AFM could be both on the nonhomogeneous surface and PMAA residues on SLG. However, the accurate thickness of SLG was calculated by SPR angle shift. The SPR angle (θ_{SPR}) shifts when SLG is in contact with the Au surface (Fig. 3.10B), and in comparison with a bare gold surface is used to calculate the thickness of SLG using Eq. 3.6.[73] The estimated thickness of SLG using SPR was 1.3, well within the AFM-based method.

$$d = \frac{l_d}{2}\left(\frac{R}{R_{max}}\right) \tag{3.6}$$

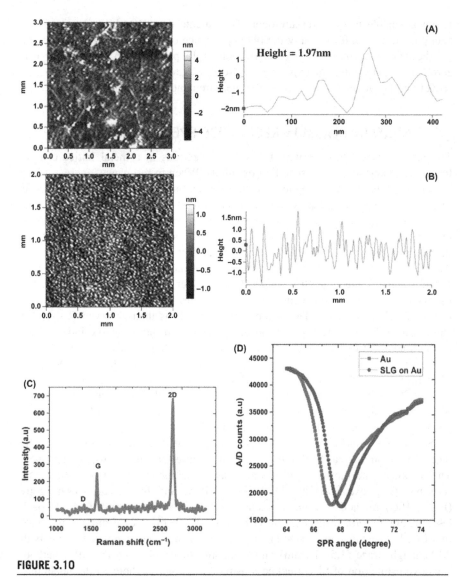

FIGURE 3.10

(A) AFM image and height profile of SLG on Au surface. (B) AFM image and height profile of a bare Au surface. (C) Raman spectra confirm the presence of SLG transferred to the Au surface. (D) Change in SPR angle of the Au surface (*left* after deposition of SLG (*right*).

d, thickness of the adsorbed material; R_{max}, SPR response for an infinitely thick layer; l_d, decay length of an evanescent wave; R, the SPR response of the film.

SPR offers an inexpensive, nondestructive approach to characterize large-area graphene films, with high sensitivity. More work has to be done to prove whether this technique can distinguish SLG, bi- or trilayer graphene samples. Graphene thin film on gold surface can be harnessed to study the interaction of various molecules with graphene at high sensitivity, which could lead to the development of graphene-based sensors. SPR data can complement microscopic techniques to calculate flake thicknesses, which are discussed next.

3.4 MICROSCOPY METHODS

Direct visualization of graphene on substrates is done using state-of-the-art microscopy instruments. Qualitative and quantitative details of the graphene flakes are recognized using electron microscopy.[74] Success and the degree of exfoliation, number of layers, lateral size, atomic-level defects, and field emission characteristics have been studied using scanning electron microscopy (SEM), transmission electron microscopy (TEM), AFM, and scanning tunneling electron microscopy (STEM). These cutting-edge methods harness atomic-level details of graphene and are discussed in the following sections in detail.

3.4.1 SEM AND TEM

High-voltage electron beams under high vacuum are used to examine graphene, which possesses high transmission to electrons.[75] Graphite flakes consisting of thousands of graphene layers in the crystallites have poor transmittance to electrons, but can be imaged by SEM (Fig. 3.11A).[8] The graphite flakes (Sigma Aldrich, +100 mesh) are several microns thick in size and the in-plane view clearly shows the graphene stacks (Fig. 3.11A). Upon exfoliation and separation of the individual layers from the stacks, thin layers with few (1−5) sheets are visualized under TEM (Fig. 3.11B).[8] This is clear evidence of successful exfoliation, which should also match with the spectroscopic investigations described in this chapter, particularly Raman fingerprints. High-resolution TEM (HR-TEM) images can be combined with the selected area electron diffraction (SAED) pattern of monolayer graphene for a better understanding of the material (Fig. 3.11C,D).[76] The diffraction pattern typically shows a sixfold symmetry, and multiple patterns appear as the number of layers in the flake increases.[77] Qualitatively, the visual inspection of the sheets is used to analyze the number of layers present, but quantitative information about polydispersity of the graphene samples has been studied by TEM.

The lateral size of graphene sheets and size distribution are quantitated by measuring the length of individual sheets by counting a statistically significant number of objects (>100). Image processing software ImageJ is an effective tool to perform size analysis of the flakes in microscopic images.[78] Well-separated (non-aggregated) flakes are essential to reduce the error in these measurements because evaporation of the graphene suspensions on the grid could induce aggregation of the flakes.[46] This can alter the size distributions, additionally, leakage

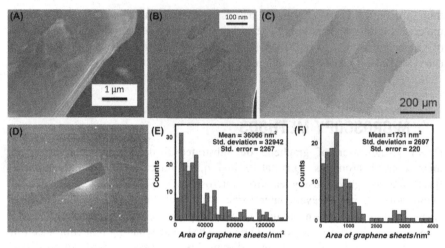

FIGURE 3.11

(A) SEM image of graphite (© John Wiley & Sons). (B) TEM image of exfoliated, few-layer graphene. (C) Monolayer of CVD graphene under HR-TEM (© Elsevier). (D) SAED pattern of monolayer graphene (© Elsevier).[76] (E, F) Area measurements of graphene sheets in a typical sample, after exfoliation and fractionation of the samples, as marked.[15]

of the smaller flakes through the lacy grids might also bias these estimates. Comparisons of TEM measurements with Raman analysis (Section 3.3.2) are essential to validate the observed lateral dimensions of any of these samples. Raman spectroscopy may not be accurate in examining very small sheets (<100 nm), present in shear-exfoliated graphene suspensions (10%−15%), but microscopy images clearly displayed this fraction (Fig. 3.6A). The smaller size fraction was separated by centrifugation at 8000 rpm for 90 minutes[10] and the samples quantified using extinction measurements as well as Raman spectroscopy (Fig. 3.5B).[15]

TEM analysis of the samples is under dry conditions and the effect of drying on exfoliated graphene samples may cause crumbling, folding, stacking, and particle aggregation which may or may not be avoided. Increasing the number of sheets analyzed, and "hand picking" separated, unfolded sheets may provide a better representation of the features of graphene suspensions.[46] TEM size statistics are usually presented as "lateral size," which is the largest dimension of any given sheet, but the product of the length and the width (area) of a two-dimensional object is more meaningful to establish than size. High-aspect (elongated) sheets are better represented in area measurements, which give higher size range in length evaluations (Fig. 3.11E). The area of the smaller, individual sheets can be measured using ImageJ software (Fig. 3.11F).[15] Again, errors due to drying effects cannot be accounted for with this estimation. Sampling of the graphene dispersions is relatively easier than CVD-grown graphene, where transfer of graphene from the substrate to TEM grid is not required. Likely errors in measurements due to sample damage, folding, and/or aggregation of the sheets during the transfer are anticipated.[79] Another surface measurement technique, AFM, is useful to examine nontransmittable surfaces, this is described next.

3.4.2 ATOMIC FORCE MICROSCOPY/SCANNING TUNNELING MICROSCOPY

AFM has an interesting relation to the discovery of graphene. Isolation of graphene from graphite crystals using Scotch tape was common practice to clean surfaces prior to imaging. Evidently, no one looked into the cleaved portion of graphite stuck to the tape, until Geim and Novoselov in 2004.[80] AFM is widely used in graphene research to characterize graphene topography, size, thickness, and atomic defects, by following the movement of a cantilever on the graphene surface, relative to that on the substrate itself.[81] Moreover, mechanical and electronic properties of graphene (composites) are characterized using sophisticated variations of the AFM techniques.[82,83] Another advantage of AFM is its capability of solution phase imaging, which is not possible in electron microscopy.[84] The measured thickness of graphene using AFM provides direct information about the number of layers present in the graphene flakes (Fig. 3.12A).

The reported thickness of single-layer graphene ranges from 0.4 to 1.7 nm,[85] but it also depends on the type of cantilever tip used, the surface composition of the graphene, the uniformity of the underlying substrate, the surface roughness of the substrate, and the quality of the image feedback. Smaller thickness of 0.1−0.3 nm was noted with a carbon nanotube-modified AFM tip.[86] The thickness measurement also depends on the interactions of the tip with the substrate, as well as the solvent, substrate, or exfoliating agent used.[86] For liquid-phase exfoliated graphene samples, histograms are generated to account for the fraction of mono-, bi-, tri- or multiple layers (Fig. 3.12B), and this measurement is complementary to Raman analysis.[40] Atomic-level imaging of the honeycomb lattice of the top-most layer of highly oriented pyrolytic graphite (HOPG) is performed by scanning tunneling microscopy (Fig. 3.12C). The image displays the atomic layer of hexagonal carbon bonds.[40] Thus, atomic and molecular modification of the graphene basal plane can be scanned using AFM or STM.

Graphene composites with biomolecules, where protein or DNA strongly adsorb to the graphene layers, can be visualized by AFM imaging and height analysis. AFM images of protein-adsorbed GO indicated islands of proteins and the height of the protein layer was comparable to its known diameter (Fig. 3.12D).[61] For example, bovine serum albumin strongly adsorbed onto GO measures $\sim$40 nm height on line scan across the plane of the sheet. This number was very close to the theoretical size ($\sim$55 nm, much larger than protein diameter), which indicated deformation of protein structure on the GO surface, and loss of protein structure on binding to GO was confirmed by circular dichroism

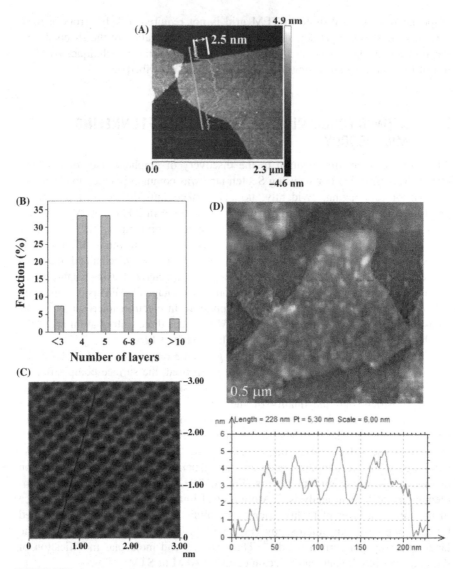

FIGURE 3.12

(A) AFM image and height measurement of few-layer graphene exfoliated in Oxone (© Elsevier).[40] (B) Histogram shows the fraction of each number of layers present in suspension[40] (© Elsevier). (C) Graphene layer in HOPG, scanned using a tunneling microscope (© Elsevier).[40] (D) Bovine serum albumin (BSA) modified with polyamine-bound GO sheets shows about 40 nm diameter in height profile analysis (bottom), closer to the theoretical size (55 nm). © American Chemical Society.[61]

spectroscopy.[61] Similarly, flattening of DNA on the graphene surface was also shown by scanning force microscopy, where double-stranded DNA showed an average height of 0.4 nm, much smaller than 2 nm (the known size in the solution phase).[87] These structure deformations would affect activities of biomolecules, but surface passivation approaches are developed to bypass direct surface interactions of the biomolecules with GO.[61]

3.4.3 FLUORESCENCE QUENCHING MICROSCOPY

Graphene derivatives are strong quenchers of fluorescence of many organic dyes via energy transfer or other mechanisms (Fig. 3.13A).[88] Fluorescence quenching by graphene is used to image by using fluorescent dyes as negative stains. The substrate to be bound to graphene is coated with a fluorescent dye which has high quantum yield and strong quenching by graphene (fluorescein, for example).[89] After the addition of the substrate to the graphene, it can be visualized by a florescence microscope (Fig. 3.13).[89] The graphene−substrate regions show a clear difference in contract, when compared to the regions of the free dye domains or unoccupied graphene domains. Interestingly, GO and rGO can be distinguished by this technique, because of their difference in texture and quenching efficiencies. Low resolution of conventional fluorescent microscopes limits the applicability of fluorescence quenching microscopy (FQM) in examining nanometer-level sheets.[90] However, application of super-resolution imaging can overcome this problem, where submicron-size sheets have been imaged.[90] Photo-degradation of the fluorescent dye molecule is a possible problem, especially over prolonged imaging.[90] FQM offers a simple and preliminary characterization of graphene, which should be backed up by data from other techniques.

3.5 CONCLUSIONS

As graphene research has been developing rapidly over time, new characterization methods to enhance the current and classical methods may be needed to ensure the quality and reproducibility of these measurements. Large-scale preparation of graphene by economic methods requires fast, reliable, robust, and low-cost characterization methods. The major characteristics to be examined depend to a large extent on the intended application of the graphene produced, but few properties must be analyzed to control and predict the properties of the preparation in any event. One of the major goals of examining graphene and its derivatives is to clearly separate the characteristics of graphene, graphene oxide, and reduced graphene oxide, which are interchangeably denoted as graphene in the literature.

One of the best ways to identify these different forms is by Raman spectroscopy. In our experience, a first glance at high-quality Raman data help in rapid identification of these specific forms. The I_D/I_G value is directly proportional to

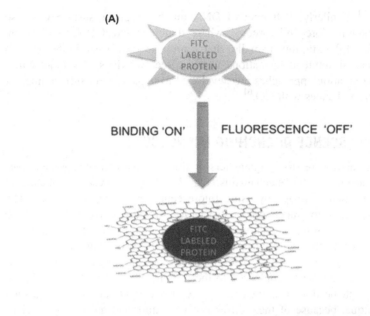

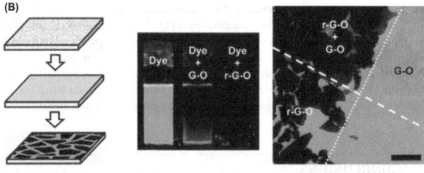

FIGURE 3.13

(A) Illustration of fluorescence quenching property of graphene oxide. Fluorescently labeled protein was readily quenched by the addition of GO (© American Chemical Society).[61] (B) Fluorescence quenching microscopy image of graphene oxide and reduced graphene oxide (scale bar on the right-hand-side image is 25 μm).[89]

the extent of defects/oxidation, but visual inspection of the spectra is of sufficient value to gather useful information.

Although, quantitative analysis of Raman data is not popular, more of its use is expected in the future because of its power in identifying the unique characteristics of single-layer graphene and to some extent few-layer graphene. Apart from spectroscopic techniques, a major microscopy image must also be made to better characterize the quality and nature of the graphene produced by a given method, prior to any further study or application of the given preparation.

TEM and AFM data are complementary to Raman analysis, which provides information about the extent of success of exfoliation, lateral size, and number of layers present in the flakes. However, the microscopy methods alone cannot distinguish graphene from its oxide derivatives or chemical derivatives unless coupled with methods such as EDX (energy dispersive x-ray spectroscopy). Depending on availability, one of these techniques is considered as a benchmark for graphene imaging and characterization, in addition to the spectroscopic methods discussed above. Also, direct imaging helps to characterize graphene composites in a complex matrix, where the counterpart systems are not accessible to spectroscopic techniques. Finally, the drying effects and interference from exfoliating agents should be considered while interpreting the data.

REFERENCES

1. Liu, W.-W.; Chai, S.-P.; Mohamed, A. R.; Hashim, U. Synthesis and Characterization of Graphene and Carbon Nanotubes: A Review on the Past and Recent Developments. *J. Ind. Eng. Chem.* **2014,** *20* (4), 1171−1185.
2. Hernandez, Y.; Nicolosi, V.; Lotya, M.; Blighe, F. M.; Sun, Z.; De, S., et al. High-Yield Production of Graphene by Liquid-Phase Exfoliation of Graphite. *Nat Nano.* **2008,** *3* (9), 563−568.
3. Backes, C.; Paton, K. R.; Hanlon, D.; Yuan, S.; Katsnelson, M. I.; Houston, J., et al. Spectroscopic Metrics Allow In Situ Measurement of Mean Size and Thickness of Liquid-Exfoliated Few-Layer Graphene Nanosheets. *Nanoscale.* **2016,** *8* (7), 4311−4323.
4. Ferrari, A. C.; Basko, D. M. Raman Spectroscopy as a Versatile Tool for Studying the Properties of Graphene. *Nat Nano.* **2013,** *8* (4), 235−246.
5. Yang, D.; Velamakanni, A.; Bozoklu, G.; Park, S.; Stoller, M.; Piner, R. D.; Stankovich, S.; Jung, I.; Field, D. A.; Ventrice, C. A., Jr; Ruoff, R. S. Chemical Analysis of Graphene Oxide Films After Heat and Chemical Treatments by X-Ray Photoelectron and Micro-Raman Spectroscopy. *Carbon* **2009,** *47* (1), 145−152.
6. Guerrero-Contreras, J.; Caballero-Briones, F. Graphene Oxide Powders With Different Oxidation Degree, Prepared by Synthesis Variations of the Hummers Method. *Mater. Chem. Phys.* **2015,** *153*, 209−220.
7. Montes-Navajas, P.; Asenjo, N. G.; Santamaría, R.; Menéndez, R.; Corma, A.; García, H. Surface Area Measurement of Graphene Oxide in Aqueous Solutions. *Langmuir.* **2013,** *29* (44), 13443−13448.
8. Pattammattel, A.; Kumar, C. V. Kitchen Chemistry 101: Multigram Production of High Quality Biographene in a Blender with Edible Proteins. *Adv. Functi. Mater.* **2015,** *25* (45), 7088−7098.
9. Investigation of Structural and Electronic Properties of Graphene Oxide. Appl. Phys. Lett. 2011;99(1):013104.
10. Varrla, E.; Paton, K. R.; Backes, C.; Harvey, A.; Smith, R. J.; McCauley, J.; Coleman, J. N. Turbulence-Assisted Shear Exfoliation of Graphene Using Household Detergent and a Kitchen Blender. *Nanoscale* **2014,** *6* (20), 11810−11819.

11. Chabot, V.; Kim, B.; Sloper, B.; Tzoganakis, C.; Yu, A. High Yield Production and Purification of Few Layer Graphene by Gum Arabic Assisted Physical Sonication. *Sci. Rep.* **2013,** *3*, 1378.

12. Ager, D.; Arjunan Vasantha, V.; Crombez, R.; Texter, J. Aqueous Graphene Dispersions—Optical Properties and Stimuli-Responsive Phase Transfer. *ACS Nano* **2014,** *8* (11), 11191−11205.

13. Lotya, M.; Hernandez, Y.; King, P. J.; Smith, R. J.; Nicolosi, V.; Karlsson, L. S.; Blighe, F. M.; De, S.; Wang, Z.; McGovern, I. T.; Duesberg, G. S.; Coleman, J. N. Liquid Phase Production of Graphene by Exfoliation of Graphite in Surfactant/Water Solutions. *J. Am. Chem. Soc.* **2009,** *131* (10), 3611−3620.

14. Uysal Unalan, I.; Wan, C.; Trabattoni, S.; Piergiovanni, L.; Farris, S. Polysaccharide-Assisted Rapid Exfoliation of Graphite Platelets Into High Quality Water-Dispersible Graphene Sheets. *RSC Adv.* **2015,** *5* (34), 26482−26490.

15. Pattammattel, A.; Pande, P.; Kuttappan, D.; Basu, A.B.; Amalaradjou, A.; Kumar, C.V. Graphite exfoliation in animal sera: Scalable Production, Enhanced Stability and Reduced Toxicity, *Submitted Manuscript.*

16. Ferrari, A. C.; Meyer, J. C.; Scardaci, V.; Casiraghi, C.; Lazzeri, M.; Mauri, F.; Piscanec, S.; Jiang, D.; Novoselov, K. S.; Roth, S.; Geim, A. K. Raman Spectrum of Graphene and Graphene Layers. *Phys. Rev. Lett.* **2006,** *97* (18), 187401.

17. Ferrari, A. C. Raman Spectroscopy of Graphene and Graphite: Disorder, Electron−Phonon Coupling, Doping and Nonadiabatic Effects. *Solid State Commun.* **2007,** *143* (1−2), 47−57.

18. hPaton, K. R.; Varrla, E.; Backes, C.; Smith, R. J.; Khan, U.; O'Neill, A.; Boland, C.; Lotya, M.; Istrate, O. M.; King, P., et al. Scalable Production of Large Quantities of Defect-Free Few-Layer Graphene by Shear Exfoliation in Liquids. *Nat. Mater.* **2014,** *13* (6), 624−630.

19. Liu, Y.; Liu, Z.; Lew, W. S.; Wang, Q. J. Temperature Dependence of the Electrical Transport Properties in Few-Layer Graphene Interconnects. *Nanoscale Res. Lett.* **2013,** *8* (1), 335.

20. Childres, I.; Jauregui, L. A.; Park, W.; Cao, H.; Chen, Y. P. Raman Spectroscopy of Graphene and Related Materials. In *New Developments in Photon and Materials Research;* Jang, J. I., Ed.; Nova Science Publishers: Hauppauge NY, 2013.

21. Yu, P. Y.; Cardona, M. *Fundamentals of Semiconductors;* Springer: Berlin, 2005.

22. Gray, D.; McCaughan, A.; Mookerji, B. Crystal Structure of Graphite, Graphene and Silicon. *Phys. Solid-State Appl.* **2009,** *6*, 730.

23. Wang, J.; Deng, S.; Liu, Z.; Liu, Z. The Rare Two-Dimensional Materials With Dirac Cones. *Nat. Sci. Rev.* **2015,** *2* (1), 22−39.

24. Parviz, D.; Das, S.; Ahmed, H. S. T.; Irin, F.; Bhattacharia, S.; Green, M. J. Dispersions of Non-Covalently Functionalized Graphene with Minimal Stabilizer. *ACS Nano* **2012,** *6* (10), 8857−8867.

25. Park, K. H.; Kim, B. H.; Song, S. H.; Kwon, J.; Kong, B. S.; Kang, K.; Jeon, S. Exfoliation of Non-Oxidized Graphene Flakes for Scalable Conductive Film. *Nano Lett.* **2012,** *12* (6), 2871−2876.

26. Shang, N. G.; Papakonstantinou, P.; Sharma, S.; Lubarsky, G.; Li, M.; McNeill, D. W.; Quinn, A. J.; Zhou, W.; Blackley, R. Controllable Selective Exfoliation of High-Quality Graphene Nanosheets and Nanodots by Ionic Liquid Assisted Grinding. *Chem. Commun.* **2012,** *48* (13), 1877−1879.

27. Hasan, T.; Torrisi, F.; Sun, Z.; Popa, D.; Nicolosi, V.; Privitera, G.; Bonaccorso, F.; Ferrari, A. C. Solution-Phase Exfoliation of Graphite for Ultrafast Photonics. *Physica Status Solidi (b)* **2010,** *247* (11-12), 2953−2957.

28. Kang, M. S.; Kim, K. T.; Lee, J. U.; Jo, W. H. Direct Exfoliation of Graphite Using a Non-Ionic Polymer Surfactant for Fabrication of Transparent and Conductive Graphene Films. *J. Mater. Chem. C* **2013,** *1* (9), 1870−1875.

29. Behabtu, N.; Lomeda, J. R.; Green, M. J.; Higginbotham, A. L.; Sinitskii, A.; Kosynkin, D. V.; Tsentalovich, D.; Parra-Vasquez, A. N. G.; Schmidt, J.; Kesselman, E.; Cohen, Y.; Talmon, Y.; Tour, J. M.; Pasquali, M. Spontaneous High-Concentration Dispersions and Liquid Crystals of Graphene. *Nat. Nano* **2010,** *5* (6), 406−411.

30. Nuvoli, D.; Valentini, L.; Alzari, V.; Scognamillo, S.; Bon, S. B.; Piccinini, M.; Illescas, J.; Mariani, A. High Concentration Few-Layer Graphene Sheets Obtained by Liquid Phase Exfoliation of Graphite in Ionic Liquid. *J. Mater. Chem.* **2011,** *21* (10), 3428−3431.

31. Lotya, M.; King, P. J.; Khan, U.; De, S.; Coleman, J. N. High-Concentration, Surfactant-Stabilized Graphene Dispersions. *ACS Nano* **2010,** *4* (6), 3155−3162.

32. Shahil, K. M. F.; Balandin, A. A. Graphene−Multilayer Graphene Nanocomposites as Highly Efficient Thermal Interface Materials. *Nano Letters* **2012,** *12* (2), 861−867.

33. Fu, W.; Kiggans, J.; Overbury, S. H.; Schwartz, V.; Liang, C. Low-Temperature Exfoliation of Multilayer-Graphene Material From FeCl3 and CH3NO2 Co-Intercalated Graphite Compound. *Chem. Commun.* **2011,** *47* (18), 5265−5267.

34. Buzaglo, M.; Shtein, M.; Kober, S.; Lovrincic, R.; Vilan, A.; Regev, O. Critical Parameters in Exfoliating Graphite Into Graphene. *Phys. Chem. Chem. Phys.* **2013,** *15* (12), 4428−4435.

35. Shih, C.-J.; Vijayaraghavan, A.; Krishnan, R.; Sharma, R.; Han, J.-H.; Ham, M.-H.; Jin, Z.; Lin, S.; Paulus, G. L. C.; Reuel, N. F.; Wang, Q. H.; Blankschtein, D.; Strano, M. S. Bi- and Trilayer Graphene Solutions. *Nat. Nano* **2011,** *6* (7), 439−445.

36. Min, Y.; Zhigang, S.; Xiaojing, Z.; Shulin, M. Achieving Concentrated Graphene Dispersions in Water/Acetone Mixtures by the Strategy of Tailoring Hansen Solubility Parameters. *J. Phys. D Appl. Phys.* **2013,** *46* (2), 025301.

37. Wang, H.; Robinson, J. T.; Li, X.; Dai, H. Solvothermal Reduction of Chemically Exfoliated Graphene Sheets. *J. Am. Chem. Soc.* **2009,** *131* (29), 9910−9911.

38. Englert, J. M.; Röhrl, J.; Schmidt, C. D.; Graupner, R.; Hundhausen, M.; Hauke, F.; Hirsch, A. Soluble Graphene: Generation of Aqueous Graphene Solutions Aided by a Perylenebisimide-Based Bolaamphiphile. *Adv. Mater.* **2009,** *21* (42), 4265−4269.

39. Safavi, A.; Tohidi, M.; Mahyari, F. A.; Shahbaazi, H. One-Pot Synthesis of Large Scale Graphene Nanosheets From Graphite-Liquid Crystal Composite Via Thermal Treatment. *J. Mater. Chem.* **2012,** *22* (9), 3825−3831.

40. Tian, S.; Yang, S.; Huang, T.; Sun, J.; Wang, H.; Pu, X.; Tian, L.; He, P.; Ding, G.; Xie, X. One-Step Fast Electrochemical Fabrication of Water-Dispersible Graphene. *Carbon* **2017,** *111*, 617−621.

41. Ge, Y.; Wang, J.; Shi, Z.; Yin, J. Gelatin-Assisted Fabrication of Water-Dispersible Graphene and Its Inorganic Analogues. *J. Mater. Chem.* **2012,** *22* (34), 17619−17624.

42. Yu, Z.; Shi, Z.; Xu, H.; Ma, X.; Tian, M.; Yin, J. Green Chemistry: Co-Assembly of Tannin-Assisted Exfoliated Low-Defect Graphene and Epoxy Natural Rubber Latex to Form Soft and Elastic Nacre-Like Film With Good Electrical Conductivity. *Carbon* **2017,** *114*, 649−660.

43. Bepete, G.; Anglaret, E.; Ortolani, L.; Morandi, V.; Huang, K.; Pénicaud, A.; Drummond, C. Surfactant-free single-layer graphene in water. *Nat. Chem.* 2016, *advance online publication*.

44. Gravagnuolo, A. M.; Morales-Narváez, E.; Longobardi, S.; da Silva, E. T.; Giardina, P.; Merkoçi, A. In Situ Production of Biofunctionalized Few-Layer Defect-Free Microsheets of Graphene. *Adv. Funct. Mater.* **2015,** *25* (18), 2771−2779.

45. Kumar, C. V.; Pattammattel, A. Chapter Eleven - BioGraphene: Direct Exfoliation of Graphite in a Kitchen Blender for Enzymology Applications. In *Methods in Enzymology;* Challa Vijaya, K., Ed.; , 571; Academic Press, 2016; pp 225−244.

46. Backes, C.; Higgins, T. M.; Kelly, A.; Boland, C.; Harvey, A.; Hanlon, D.; Coleman, J. N. Guidelines for Exfoliation, Characterization and Processing of Layered Materials Produced by Liquid Exfoliation. *Chem. Mater.* **2017,** *29* (1), 243−255.

47. Basov, D. N.; Fogler, M. M.; Lanzara, A.; Wang, F.; Zhang, Y. Colloquium. *Rev. Modern Phys.* **2014,** *86* (3), 959−994.

48. Ramos-Galicia, L.; Mendez, L. N.; Martínez-Hernández, A. L.; Espindola-Gonzalez, A.; Galindo-Esquivel, I. R.; Fuentes-Ramirez, R.; Velasco-Santos, C. Improved Performance of an Epoxy Matrix as a Result of Combining Graphene Oxide and Reduced Graphene. *Inter. J. Polymer Sci.* **2013,** *2013*, 7.

49. Park, J. S.; Cho, S. M.; Kim, W.-J.; Park, J.; Yoo, P. J. Fabrication of Graphene Thin Films Based on Layer-by-Layer Self-Assembly of Functionalized Graphene Nanosheets. *ACS Appl. Mater. Interf.* **2011,** *3* (2), 360−368.

50. Layek, R. K.; Nandi, A. K. A Review on Synthesis and Properties of Polymer Functionalized Graphene. *Polymer* **2013,** *54* (19), 5087−5103.

51. Goods, J. B.; Sydlik, S. A.; Walish, J. J.; Swager, T. M. Phosphate Functionalized Graphene With Tunable Mechanical Properties. *Adv. Mater.* **2014,** *26* (5), 718−723.

52. Yu, S.; Liu, J.; Zhu, W.; Hu, Z.-T.; Lim, T.-T.; Yan, X. Facile Room-Temperature Synthesis of Carboxylated Graphene Oxide-Copper Sulfide Nanocomposite With High Photodegradation and Disinfection Activities Under Solar Light Irradiation. *Sci. Rep.* **2015,** *5*, 16369.

53. Mallakpour, S.; Abdolmaleki, A.; Borandeh, S. Covalently Functionalized Graphene Sheets With Biocompatible Natural Amino Acids. *Appl. Surf. Sci.* **2014,** *307*, 533−542.

54. Stobinski, L.; Lesiak, B.; Malolepszy, A.; Mazurkiewicz, M.; Mierzwa, B.; Zemek, J.; Jiricek, P.; Bieloshapka, I. Graphene Oxide and Reduced Graphene Oxide Studied by the XRD, TEM and Electron Spectroscopy Methods. *J. Elect. Spectros. Related Phenomena* **2014,** *195*, 145−154.

55. Susi, T.; Pichler, T.; Ayala, P. X-Ray Photoelectron Spectroscopy of Graphitic Carbon Nanomaterials Doped With Heteroatoms. *Beilstein J. Nanotechnol.* **2015,** *6*, 177−192.

56. Scardamaglia, M.; Aleman, B.; Amati, M.; Ewels, C.; Pochet, P.; Reckinger, N.; Colomer, J. F.; Skaltsas, T.; Tagmatarchis, N.; Snyders, R.; Gregoratti, L.; Bittencourt, C. Nitrogen Implantation of Suspended Graphene Flakes: Annealing Effects and Selectivity of sp^2 Nitrogen Species. *Carbon* **2014,** *73*, 371−381.

57. Thermoscientific X.P.S. data base, Carbon, http://xpssimplified.com/elements/carbon.php.

58. Cooper, A. J.; Wilson, N. R.; Kinloch, I. A.; Dryfe, R. A. W. Single Stage Electrochemical Exfoliation Method for the Production of Few-Layer Graphene Via Intercalation of Tetraalkylammonium Cations. *Carbon* **2014,** *66*, 340−350.

59. Yi, M.; Shen, Z. Kitchen Blender for Producing High-Quality Few-Layer Graphene. *Carbon* **2014,** *78,* 622−626.

60. Bacon, G. The Interlayer Spacing of Graphite. *Acta Crystallograp.* **1951,** *4* (6), 558−561.

61. Pattammattel, A.; Puglia, M.; Chakraborty, S.; Deshapriya, I. K.; Dutta, P. K.; Kumar, C. V. Tuning the Activities and Structures of Enzymes Bound to Graphene Oxide with a Protein Glue. *Langmuir* **2013,** *29* (50), 15643−15654.

62. Dikin, D. A.; Stankovich, S.; Zimney, E. J.; Piner, R. D.; Dommett, G. H. B.; Evmenenko, G.; Nguyen, S. T.; Ruoff, R. S. Preparation and Characterization of Graphene Oxide Paper. *Nature* **2007,** *448* (7152), 457−460.

63. Kovtyukhova, N. I.; Wang, Y.; Berkdemir, A.; Cruz-Silva, R.; Terrones, M.; Crespi, V. H.; Mallouk, T. E. Non-Oxidative Intercalation and Exfoliation of Graphite by Brønsted Acids. *Nat. Chem.* **2014,** *6* (11), 957−963.

64. Lyklema, J. *Fundamentals of Interface and Colloid Science: Soft Colloids*, 5. Academic Press: San Diego, 2005.

65. LykBooth, F. Theory of Electrokinetic Effects. *Nature* **1948,** *161* (4081), 83−86.

66. American Society for Testing and Materials. ASTM Standard D 4187-8, Zeta Potential of Colloids in Water and Waste Water, 1985.

67. Kashyap, S.; Mishra, S.; Behera, S. K. Aqueous Colloidal Stability of Graphene Oxide and Chemically Converted Graphene. *J. Nanopart.* **2014,** *2014,* 6.

68. Ahadian, S.; Estili, M.; Surya, V. J.; Ramon-Azcon, J.; Liang, X.; Shiku, H.; Ramalingam, M.; Matsue, T.; Sakka, Y., et al. Facile and Green Production of Aqueous Graphene Dispersions for Biomedical Applications. *Nanoscale* **2015,** *7* (15), 6436−6443.

69. Zeng, S.; Baillargeat, D.; Ho, H.-P.; Yong, K.-T. Nanomaterials Enhanced Surface Plasmon Resonance for Biological and Chemical Sensing Applications. *Chem. Soc. Rev.* **2014,** *43* (10), 3426−3452.

70. Jussila, H.; Yang, H.; Granqvist, N.; Sun, Z. Surface Plasmon Resonance for Characterization of Large-Area Atomic-Layer Graphene Film. *Optica* **2016,** *3* (2), 151−158.

71. Unpublished results from our laboratories, 2017.

72. Park, H.; Brown, P. R.; Bulovice, V.; Kong, J. Graphene as Transparent Conducting Electrodes in Organic Photovoltaics: Studies in Graphene Morphology, Hole Transporting Layers, and Counter Electrodes. *Nano Lett.* **2011,** *12* (1), 133-−1140.

73. Jung, L. S.; Campbell, C. T.; Chinowsky, T. M.; Mar, M. N.; Yee, S. S. Quantitative Interpretation of the Response of Surface Plasmon Resonance Sensors to Adsorbed Films. *Langmuir* **1998,** *14* (19), 5636−5648.

74. Allen, M. J.; Tung, V. C.; Kaner, R. B. Honeycomb Carbon: A Review of Graphene. *Chem. Rev.* **2010,** *110* (1), 132−145.

75. Wilson, N. R.; Pandey, P. A.; Beanland, R.; Young, R. J.; Kinloch, I. A.; Gong, L.; Liu, Z.; Suenaga, K.; Rourke, J. P.; York, S. J.; Sloan, J. Graphene Oxide: Structural Analysis and Application as a Highly Transparent Support for Electron Microscopy. *ACS Nano* **2009,** *3* (9), 2547−2556.

76. Song, Y.; Pan, D.; Cheng, Y.; Wang, P.; Zhao, P.; Wang, H. Growth of Large Graphene Single Crystal Inside a Restricted Chamber by Chemical Vapor Deposition. *Carbon* **2015,** *95,* 1027−1032.

77. Zan, R.; Ramasse, Q.M.; Jalil, R.; Bangert, U. Atomic Structure of Graphene and h-BN Layers and Their Interactions with Metals, 2013.

78. Schneider, C. A.; Rasband, W. S.; Eliceiri, K. W. NIH Image to ImageJ: 25 Years of Image Analysis. *Nat. Methods* **2012,** *9* (7), 671−675.

79. Li, B.; Nan, Y.; Zhang, P.; Song, X. Structural Characterization of Individual Graphene Sheets Formed by Arc Discharge and Their Growth Mechanisms. *RSC Adv.* **2016,** *6* (24), 19797−19806.

80. Andrei, E. Y.; Li, G.; Du, X. Electronic Properties of Graphene: A Perspective From Scanning Tunneling Microscopy and Magnetotransport Rep. *Prog. Phys.* **2012,** *75,* 056501.

81. Paredes, J. I.; Villar-Rodil, S.; Solís-Fernández, P.; Martínez-Alonso, A.; Tascón, J. M. D. Atomic Force and Scanning Tunneling Microscopy Imaging of Graphene Nanosheets Derived from Graphite Oxide. *Langmuir* **2009,** *25* (10), 5957−5968.

82. Burnett, T.; Yakimova, R.; Kazakova, O. Mapping of Local Electrical Properties in Epitaxial Graphene Using Electrostatic Force Microscopy. *Nano Lett.* **2011,** *11* (6), 2324−2328.

83. Cao, G. Atomistic Studies of Mechanical Properties of Graphene. *Polymers* **2014,** *6* (9), 2404.

84. Park, J.; Adiga, V. P.; Zettl, A.; Paul Alivisatos, A. High Resolution Imaging in the Graphene Liquid Cell. In *Liquid Cell Electron Microscopy;* Ross, F. M., Ed.; Cambridge University Press: Cambridge, 2016; pp 391−392.

85. Nemes-Incze, P.; Osváth, Z.; Kamarás, K.; Biró, L. P. Anomalies in Thickness Measurements of Graphene and Few Layer Graphite Crystals by Tapping Mode Atomic Force Microscopy. *Carbon* **2008,** *46* (11), 1435−1442.

86. Cameron, J. S.; Ashley, D. S.; Andrew, J. S.; Joseph, G. S.; Christopher, T. G. Accurate Thickness Measurement of Graphene. *Nanotechnology* **2016,** *27* (12), 125704.

87. Severin, N.; Dorn, M.; Kalachev, A.; Rabe, J. P. Replication of Single Macromolecules with Graphene. *Nano Lett.* **2011,** *11* (6), 2436−2439.

88. Kasry, A.; Ardakani, A. A.; Tulevski, G. S.; Menges, B.; Copel, M.; Vyklicky, L. Highly Efficient Fluorescence Quenching With Graphene. *J. Phys. Chem. C* **2012,** *116* (4), 2858−2862.

89. Kim, J.; Cote, L. J.; Kim, F.; Huang, J. Visualizing Graphene Based Sheets by Fluorescence Quenching Microscopy. *J. Am. Chem. Soc.* **2010,** *132* (1), 260−267.

90. Stöhr, R. J.; Kolesov, R.; Xia, K.; Reuter, R.; Meijer, J.; Logvenov, G.; Wrachtrup, J. Super-resolution Fluorescence Quenching Microscopy of Graphene. *ACS Nano* **2012,** *6* (10), 9175−9181.

Inorganic analogues of graphene: synthesis, characterization, and applications

CHAPTER OUTLINE

4.1 Summary 75
4.2 Introduction 76
4.3 Hexagonal Boron Nitride 77
    4.3.1 Production 78
    4.3.2 Characterization Techniques 79
    4.3.3 Applications of h-BN 82
4.4 Transition Metal Dichalcogenides 83
    4.4.1 Exfoliation 83
    4.4.2 Characterization Techniques 85
    4.4.3 Applications 86
4.5 MXenes 87
    4.5.1 Production 88
    4.5.2 Characterization Techniques 90
    4.5.3 Applications 91
4.6 Monoatomic Layers 91
    4.6.1 Synthesis or Exfoliation 92
    4.6.2 Characterization 93
    4.6.3 Applications 94
4.7 Conclusions 94
References 95

4.1 SUMMARY

The structure, synthesis, characterization, and applications of inorganic analogues of graphene (two-dimensional (2-D) materials) are discussed in this chapter. Hexagonal boron nitride (h-BN), also known as "white graphene" is discussed first. Exfoliation of h-BN in organic solvents and aqueous solutions are compared. Characterization of nanolayers of h-BN by specific spectroscopic and microscopic techniques is discussed from an experimental point of view. Applications of white graphene in electronics and beyond are also reviewed. Next, we discuss the transition metal dichalcogenides (TMDs), which have a graphene-like structure. MoS_2 was taken as a representative member of this family in order to discuss the preparation, characterization, and applications of TMDs. Subsequently, MXenes, which are structural analogues of TMDs, were reviewed. MXenes are being explored for applications in energy storage, electromagnetic shielding, and biomedicine. We close this chapter with the discussion of other monoatomic-layered materials such as phosphorene. We have not included metal oxides and phosphates, many of which form 2-D materials as well. This chapter offers readers a general introduction to 2-D materials other than graphene and familiarizes them with its inorganic counterparts and their interesting chemistry.

4.2 INTRODUCTION

The discovery of graphene paved the way for the isolation and synthesis of a family of materials, known as 2-D materials, or the graphene family of materials. Because of their unprecedented optoelectronic and mechanical properties, when compared to the corresponding bulk materials, the 2-D graphene-analogues have brought light to several surprises. Major attention was drawn to layered boron nitride (which is isoelectronic to graphene), TMDs (e.g., MoS_2), and MXenes. Recent approaches to create monoatomic layers of phosphorus, boron, silicon, etc. are highly promising for utilization in potential applications. Similar to the bonding in graphene, inorganic layered materials also possess a primary hexagonal building block of one or two atoms thick, which extend in the 2-D network. The van der Waals interactions between the layers stabilizes these materials in the bulk and through the use of suitable conditions, as in the case of graphene, these materials are either exfoliated from their bulk phases or synthesized from molecular precursors, as in the case of chemical vapor deposition (CVD).

Freestanding h-BN was discovered in 2005, along with the introduction of dichalcogenides.[1] In that study, Geim and coworkers isolated the monolayers of h-BN, MoS_2, $NbSe_2$, $Bi_2Sr_2CaCu_2O_x$, and graphite by simply peeling the layers off the clean bulk crystals. To date, numerous atomically thin nanosheets of various 2D bulk materials are known, which are exfoliated in solution, deposited on surfaces, or chemically synthesized.

Inspired by graphite exfoliation to produce graphene monolayers, other 2-D materials are also exfoliated from the corresponding bulk crystals by applying mechanical force to the crystal, under favorable solvent/solute conditions.[2] Exfoliation of h-BN, MoS_2, $MoSe_2$, and many others by ultrasonication, high-speed shearing, or ball milling have been carried out, although the yields, efficiencies, degree of defects, as well as the number of layers in the flakes vary from method to method.

In contrast to the above, MXenes are synthesized by etching using strong acids (HF), where M is a transition metal (Ti, V, Cr, etc.), and X represents carbon and/or nitrogen. Single-atom-thick layers of phosphorus (phosphorene) are also exfoliated from bulk black phosphorus, but other monoatomic-layer materials are synthesized by CVD. Synthetic approaches for each of these materials will be introduced in the following sections. Similar to the discussion about graphite exfoliation in the previous chapters, the liquid-phase exfoliation methods yield larger amounts of these materials with variable size and number of layers, and the CVD method is very good for the production of small amounts of high-quality materials with high precision in their atomic structure and large flake sizes. Characterization of the 2-D nanolayers is an important task where a combination of spectroscopic and microscopic tools is often used, following the path of graphene characterization.

The flake size (width and length), the number of layers, and the degree of defects are important attributes to be established by well-designed characterization methods. For exfoliated samples, the success of exfoliation is also an important screening step before producing large amounts of the sample. Raman spectroscopy is one of the most widely used techniques for the complete characterization of these materials because of its ease of implementation, precision and accuracy of the measurements, as well as well-established assignments of the observed transitions to specific features of the 2-D materials.

Quantitative measurements to analyze the properties of these nanosheets by Raman spectroscopy are not yet fully developed for these materials, when compared to those of graphene. X-ray photoelectron spectroscopy is the next popular technique to analyze these layers where the defects, doping, or impurities are readily identified and quantified. Electron microscopy and atomic force microscopy (AFM) images are important tools to examine the morphology, size, size distribution, and thickness. We will be specifically discussing multiple characterization methods for each of the materials in the next section.

4.3 HEXAGONAL BORON NITRIDE

Covalently bonded sp^2 hybridized boron and nitrogen atoms make up the h-BN structure (Fig. 4.1A), which is isoelectronic to bonding in graphene (eight valance electrons).[3] It is a very good thermal conductor (30−225 W/mK depending on pressing direction),[4] and electrical insulator (volume resistivity $>10^{14}$)[4] with a high mechanical strength (Young's modulus = 20−100 MPa).[4] Similar to graphene, it has a negative coefficient of thermal expansion ($1-4 \times 10^{-6}/°C$).[4] It is often used as a high-temperature lubricant as it is stable up to 2600°C, where most other lubricants fail.[5] h-BN is used extensively in skin care products, particularly in lipsticks and makeup materials, and is used to produce the sheen in pencils.

In bulk, BN exists in different phases such as cubic, hexagonal, amorphous, and wurtzite structures, which are analogous to the lattice of diamond, graphite,

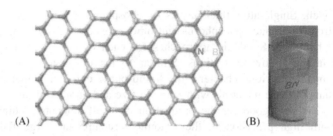

FIGURE 4.1

(A) Structure of monolayer h-BN. (B) "Milky" solution of exfoliated h-BN in water (3 mg/mL BSA, (h-BN) = 5 mg/mL, 200 mL water in a kitchen blender at 17,000 rpm speed).

amorphous carbon, and lonsdaleite (also called hexagonal diamond). Because of the structural similarity of h-BN to graphite and white texture, it is called "white graphene," whereas graphite is black. A new mineral with BN has been identified recently, but commercially it is synthesized from boron trioxide (B_2O_3) or boric acid (H_3BO_3) and ammonia (NH_3) or urea. Thermal treatment of the resulting amorphous solid yields h-BN. Individual layers of B_3N_3 nanosheets are stacked due to favorable interlayer van der Waals interactions, with d-spacings of 3.3 Å.[6] The free-standing form of h-BN of ~3.4 Å thickness can be isolated from the bulk material by mechanical and chemical processes.[7]

4.3.1 PRODUCTION

Atomically precise layers of h-BN are prepared by CVD from the decomposition of ammonia borane (BH_3NH_3) on metal surfaces.[8] However, liquid-phase exfoliation from bulk B_3N_3 is an efficient method for the production of h-BN layers, which are quite suitable for many applications, including applications in biology.[9] As discussed with graphene exfoliation (Section 2.4), solvent or suitable exfoliating agents play a major role in determining the efficiency and quality of the resulting h-BN layers. Theoretical calculations showed that polar solvents stabilize h-BN by electrostatic interactions, and solvents such as dimethyl sulfoxide and isopropyl alcohol are excellent dispersing agents to make h-BN nanolayers.[10] The Hansen solubility parameter was used as a tool to predict suitable solvent systems for the solubilization of h-BN layers. Studies used a low-boiling solvent system made of an ethanol and water mixture (45:55), which yielded exfoliation efficiencies of 0.001 mg/mL/hour through the use of ultrasonication.[11] Whereas, the use of ethanol or water alone did not exfoliate h-BN, under the same conditions, establishing the role of solvent in the mechanism of exfoliation. Besides ultrasonication, high-speed shearing,[12] vortex fluidic flow,[13] and ball milling[14] were also used successfully to delaminate the h-BN layers. Although boron nitride is less hydrophobic than graphene,[15] water by itself cannot solubilize h-BN nanolayers.[16]

h-BN nanolayers are prone to dissociation and decomposition to boron oxide and ammonia under aqueous conditions, which is enhanced by ultrasonication.[16,17] Thus, the exfoliation by sonication would lead to sheets with small lateral size because of hydrolysis at the reactive edges. Evidently, the presence of ammonia in the exfoliated mixture as well as the presence of hydroxyl groups on the edges support the hydrolysis pathway.[16] Ammonia was formed by the hydrolysis of a $B = N$ bond and replaced by $-OH$ from water molecules. Exfoliation of *h*-BN in aqueous media was achieved using surfactants, proteins, polymers, or other small molecules as stabilizing or solubilizing agents. Table 4.1 summarizes the recent data.

h-BN was also exfoliated in water using polyethylene glycol (PEG) and gelatin and the resulting product may be a good candidate for biological applications. In addition, the gelatin-based material showed the highest exfoliation efficiency among all the reported methods.[21] As shown in Table 4.1, exfoliation efficiencies of these methods are far below that reported for graphite exfoliation to graphene (Section 2.4). It is possible that shear exfoliation instead of ultrasonication could improve these efficiencies but quantitative data are not reported.[12] However, we did not find any report on shear exfoliation of boron nitride in water. The quality of *h*-BN exfoliated sheets is assessed by absorbance, Raman, XPS, TEM, and AFM characterization methods.

4.3.2 CHARACTERIZATION TECHNIQUES

When the bulk boron nitride is exfoliated successfully, one notices a "milky" white suspension that does not settle down. This is a good qualitative method to evaluate the success of the procedure (Fig. 4.1B). In addition, shining a red laser light beam through the particle suspensions results in the Tyndall effect, which is often presented in many reports.[9] The light scattering is often quantitated to measure the concentration of *h*-BN in suspension, after appropriate calibrations. *h*-BN exfoliated using PEG indicated an extinction coefficient of 4 mL/mg.m in water,[21] and this value is several-fold less than the extinction coefficient of graphene (Section 3.2.1). Absorption spectra of pure, few-layer *h*-BN, synthesized by CVD method shows a strong band at 203 nm, supported on a quartz plate (Fig. 4.2A).[23] According to the Tauc equation, this absorption corresponds to a band gap of ~ 5.6 eV.[23] Another tool to confirm success of exfoliation is by Raman spectroscopy, where the exfoliated *h*-BN shows significant changes from that of the bulk material.

Raman active mode E_{2g} of *h*-BN (similar to the G band in the Raman spectrum of graphene) is sensitive to delamination of the layers (Fig. 4.2B), and it appears around 1366 cm^{-1}.[24] In bulk material, this mode is much more intense and drops in few-layered material. There is about a 10-fold decrease in the intensity of this mode for the monolayer *h*-BN, compared to the bulk.[24] Interestingly, the peak position is also sensitive to the change in the number of layers, where monolayer *h*-BN peak occurs at higher wavenumbers (2−4 cm^{-1}) and bilayer shifts to lower wavenumbers (1−2 cm^{-1}) with respect to the bulk signal.[24] The peak width (FWHM) is

Table 4.1 Examples of Aqueous Exfoliation of *h*-BN

Method	Exfoliating Agent	Exfoliation Efficiency (mg/mL/h)	Lateral Size (μm)	Number of Layers
Ultrasonication	None[16]	4.2×10^{-3}	<1	<30 (AFM)
Ultrasonication + heating at 80°C	Poly (sodium 4-styrenesulfonate)[18]	0.02	~0.1–3	<6 (AFM)
Ultrasonication	Polyvinylpyrrolidone (M_W 10,000 g/mol)[19]	~0.38	~0.7	2–5 (TEM)
Heating at 1000–1400°C + ultrasonication	Ammonium hydrogen carbonate + ethanol/water mixture (45:55)[20]	NA	0.5–3	<6 (AFM)
Ultrasonication	Gelatin[21]	0.8	1–2	2–3 (AFM)
Heating for 4–6 days at 160–180°C	Polyethylene glycol (PEG M_r: 1500 Da)[22]	NA	0.3–0.5	3–20 (TEM)

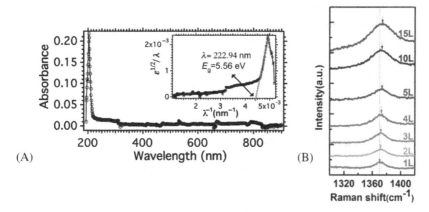

FIGURE 4.2

(A) Extinction spectra of *h*-BN monolayers on a quartz substrate (©American Chemical Society).[23] (B) Raman spectra of *h*-BN nanolayers and dependence on sheet thickness (1–3 layers [1L, 2L, and 3L]) ©American Chemical Society.[24]

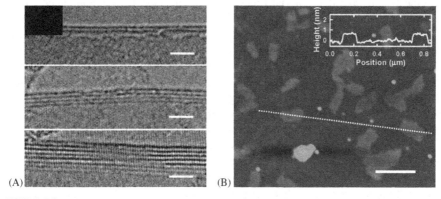

FIGURE 4.3

(A) Cross-sectional view of HR-TEM images of exfoliated *h*-BN layers (scale bar: 2 nm). (B) AFM images of *h*-BN were used to measure the thickness of exfoliated sheets (scale bar: 200 nm) with height profile shown in the inset ©American Chemical Society.[16]

insensitive to the number of layers.[24] Noticeably, Raman spectroscopy can only distinguish mono-, bi- or trilayer *h*-BN from the bulk and most of the above-listed exfoliation methods (Table 4.1) result in a mixture of bi- and trilayer materials along with multilayer *h*-BN (> 5). Thus, microscopy techniques need to be carried out to characterize the flake thickness of the exfoliated *h*-BN.

High-resolution transmission electron microscopy (HR TEM) and AFM are widely used to quantify the width/length (size) and the thickness of *h*-BN nanolayers (Fig. 4.3).[9] These measurements are done with the aid of advanced image processing software (ImageJ). Exfoliated *h*-BN sheets crumple on TEM grids, as other 2-D materials, and the lateral sizes (largest dimension) are often underestimated.[19] The number of layers in the flake is calculated by following the height profile across the edges of the flake by AFM. The expected spacing between individual layers in bulk h-BN of 0.33 nm is used to calculate the number of layers in the exfoliated samples (Fig. 4.3B).[9] This approach, however, may have errors due to the presence of exfoliating agent on the flakes. Well-characterized nanolayers of h-BN nanosheets can be used for several potential applications.

4.3.3 APPLICATIONS OF *H*-BN

The h-BN nanosheets are used in polymer composites to improve strength and decrease gas permeation,[25–27] as dielectric substrates,[28,29] in catalysis,[30] for sensing,[9,31] and for applications in the biomedical field,[32] etc. h-BN nanolayers are widely used to grow graphene via the CVD method, because of h-BN's low lattice mismatch with graphene. That is, the hexagonal lattice of h-BN drives the ordered hexagonal arrangement of carbon atoms on its surface to produce a high-quality graphene layer. The vast number of applications of graphene/h-BN composite films will be discussed in the next chapter. Boron nitride powder was previously used as filler material for polymer composites, but the nanolayer h-BN helps in reducing the amount and improves the mechanical and optical properties. h-BN sheets increase the glass transition temperature (T_g: temperature at which polymer changes from hard, glassy state to amorphous), by reducing the inter-chain mobility.[27] For example, a 22% increase in elastic modulus and an 11% increase in strength of poly(methyl methacrylate) were obtained by incorporating 0.3% (w/w) few-layer h-BN into the polymer matrix.[33] High surface area per gram of exfoliated h-BN is used to support vast numbers of nanoparticles and small molecules in catalysis, sensing, and drug-delivery applications.[9]

The synergetic effects of h-BN nanosheets and noble metal catalysts were used for oxidation reduction reactions.[9] For instance, Pt nanoparticles were integrated onto h-BN nanosheets by in situ thermal reduction and used for CO to CO_2 conversion, potentially useful in the remediation of automobile emissions.[30] Similarly, Ag nanoparticles supported on h-BN nanosheets were used in surface-enhanced Raman spectroscopy to detect very low levels of Rhodamine 6G, and more importantly the surface was regenerated after removal of the analyte by thermal annealing (400°C).[31] Highly water-dispersed, hydroxylated h-BN nanosheets were used for the delivery of doxorubicin (an anticancer drug), achieving about 300% drug loading (w/w) and ~30% release of the drug at pH 7.4.[34] The drug-loaded h-BN nanosheets were effective in dropping the cell viability to about 15% for human prostate cancerous cells (LNCaP), whereas the free drug under similar conditions dropped by only ~40% cell viability. Free h-BN had no effect on cell viability in the absence of the drug.[34] Materials such as h-BN nanosheets are complementary to graphene and they have promising applications in diverse fields. In this vein of development, structurally tunable materials such as TMDs are another remarkable class of new 2-D materials being examined.

FIGURE 4.4

Structures of TMD with general formula MX_2 (©IOP Publishing).[38]

4.4 TRANSITION METAL DICHALCOGENIDES

TMDs have the general formula of MX_2 (M, metal; X, chalcogen) and have a zig-zag X−M−X bonding, where the chalcogen has out-of-plane orientation with respect to the metal plane (Fig. 4.4).[35] Oxides, sulfides, selenides, and tellurides of transition metals such as molybdenum (Mo) and tungsten (W) form layered structures, which are similar to that of graphite. An extensive variety of TMDs are known to exist in monolayer form and one of the most widely studied members is molybdenum sulfide (MoS_2). This is because of its availability in bulk form and its band gap of 1.2−1.9 eV is comparable to that of many semiconductors, but the gap depends on the number of layers present in the flake.[36] TMD monolayers can be synthesized by CVD, produced by exfoliation from the bulk or by synthesis via the molecular beam epitaxy techniques.[37] TMDs are an interesting class of materials because of their tunable band gap as semiconductors, and their chemical composition can also be varied over a broad range.

4.4.1 EXFOLIATION

Exfoliation of TMDs from the corresponding bulk material is achieved in organic solvents, surfactants, ionic liquids, and in the presence of a number of biomolecules.[37] MoS_2 is a comprehensively studied member of this group and it is taken here as a representative example to discuss the exfoliation of TMDs, in general. Among organic solvents N-methyl pyrrolidone (NMP) was found to be the best solvent for MoS_2 exfoliation.[39] However, the presence of a trace amount of water in the solvent was necessary for success.[40] Surprisingly, the hydroperoxide form of NMP forms under sonication conditions and is the active component of exfoliation.[39] However, nonsonication processes, such as the high shear method, were also successful in obtaining stable suspensions of MoS_2.[41] Aqueous exfoliation of MoS_2, with or without stabilizing agents, has also been done.[42−44]

Table 4.2 Examples of Aqueous Exfoliation of MoS_2

Method	Exfoliating Agent	Exfoliation Efficiency (mg/mL/h)	Lateral Size (µm)	Number of Layers
Ultrasonication	Sodium cholate[43]	0.3	<1	2–9 (TEM)
Mechanical exfoliation + ultrasonication	None[44]	0.14	~0.2	4–12 (AFM)
Ball Milling + ultrasonication	SDS (sodium dodecyl sulfonate)[47]	0.06	0.05–0.7	N.A.
Ultrasonication	Tween80 (surfactant)[48]	0.3	0.05–0.3	2–16 (AFM)
Electrochemical	0.5 M Na_2SO_4[49]	NA	5–50	2–5 (AFM)
Ultrasonication	Lignin[50]	0.02	0.1–0.5 (hydrodynamic)	3–5 (TEM)
Ultrasonication	Thioglycolic acid[51]	~2	0.1–0.5	1 (AFM)
Ultrasonication	Bovine serum albumin[52]	0.04	NA	1–4 (AFM)
Ultrasonication	Silk fibroin[53]	0.5	~0.2	<4 (AFM, Raman)
Shear + turbulence	Sodium cholate[41]	0.2	0.04–0.22	2–12 (AFM)

Exfoliation of 2-D materials in water requires stabilizing agents to prevent aggregation of flakes back to the crystallites. A few examples of such exfoliation methods are compared in Table 4.2. Surfactant exfoliated MoS_2 sheets made through sonication usually result in poor efficiencies (<0.3 mg/mL/hour).[43,47,48] Thioglycolic acid was found to be one of the best exfoliation agents for MoS_2 with about 2 mg/mL/hour efficiency.[51] Biomolecules such as lignin, silk fibroin, and bovine serum albumin exfoliations yield efficiencies from 0.02 to 0.5 mg/mL/hour, which are promising for biological applications.[50,52,53] High-yield production of MoS_2 in a kitchen blender was also achieved recently, which is superior to less efficient sonication approaches.[41] In general, most of these methods harness favorable interactions between the MoS_2 layers and the exfoliation agents. But exfoliation in pure water by controlling the temperature of sonication has been reported recently.[45] However, novel methods of production are required to achieve higher concentrations of MoS_2 dispersions in water, with efficiencies comparable to those achieved with graphene.[46]

Other members of TMDs are also produced by exfoliation methods, but are usually presented as an extension of an MoS_2-based technique or as proof-of-principle studies. Noticeable members exfoliated by such methods include WS_2,[41,45]

$MoSe_2$,[45] WSe_2,[44] MoO_2,[54] TaS_2,[55] and SnS_2.[55] Characterization of these materials is primarily done by absorption, Raman, electron microscopy, and AFM methods.

4.4.2 CHARACTERIZATION TECHNIQUES

Extinction spectra of MoS_2 and other TMDs show characteristic bands in the visible region of the spectrum. For example, absorption spectra of sodium cholate exfoliated MoS_2 showed peaks around 345, 600 (λ_A), and 675 (λ_B) nm, but the peak positions are sensitive to thickness and sheet size (Fig. 4.5A).[56] The peak positions were used to quantitate the size and thickness of the sheets, following Eq. 4.1, within $\pm$ 10% accuracy[56]

$$Length, L(\mu m) = \frac{\varepsilon_{\lambda_B}/\varepsilon_{345} - 0.14}{11.5 - \varepsilon_{\lambda_B}/\varepsilon_{345}} \; ; 70 < L < 350\,nm \qquad (4.1)$$

where ε_{λ_B} is the extinction at λ_B ($\sim$ 675 nm) and ε_{345} is the extinction at 345 nm.

The number of layers present in the exfoliated MoS_2 flake (N) was calculated using the position of λ_A (600 nm) from Eq. 4.2.[56]

$$N = 2.3 \times 10^{36} e^{-54888/\lambda_A}; \; N < 10 \qquad (4.2)$$

These matrices are useful to quantitate the size and thickness of nanosheets in solution inexpensively and quickly. The absorption or scattering from the exfoliating agent or solvent must be accounted for, before applying these empirical equations to specific cases. However, with the support of other techniques, such as Raman spectroscopy and microscopy, the data can be presented.

Raman spectra of bulk MoS_2 have two prominent peaks at $\sim$ 408 cm^{-1} (A_{1g}—out-of-plane mode) and at $\sim$ 383 cm^{-1} (E_{2g}^1—in-plane mode).[57] The out-of-plane and in-plane modes correspond to the vibrations of metal-sulfur bonds.[57,58] In monolayer MoS_2, the A_{1g} mode shifts to lower wavenumbers and the E_{2g}^1 mode

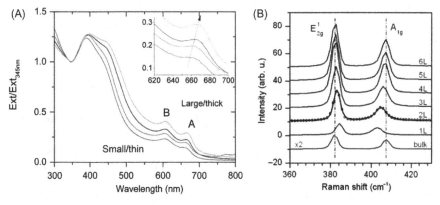

FIGURE 4.5

(A) Thickness-dependent extinction spectra (© John Wiley & Sons).[56] (B) Raman spectra of few-layer MoS_2 with a variable number of layers (©Nature Publishing group).[57]

moves to higher wavenumbers (Fig. 4.5B). However, Raman spectra of bilayer MoS_2 show no change for the position of the E_{2g}^1 mode (~ 383 cm^{-1}) but the A_{1g} mode moves to lower wavenumbers by about 1 cm^{-1}. The changes in Raman spectra are indistinguishable for the trilayer MoS_2 and the higher number of sheets from that of the bulk material.[57] Analogously, other TMDs also show changes in their band gap position or intensity changes upon exfoliation and these data have been used to distinguish them from the bulk material.[59] Microscopy studies are then required for further characterization of the nanosheets. Like other 2-D materials, the TEM and AFM images are useful to calculate the lateral size and thickness (or number of layers) of exfoliated MoS_2 sheets (Table 4.2). Next, potential applications of TMDs are discussed.

4.4.3 APPLICATIONS

Due to their direct band gap, TMDs warrant applications in electronic devices. For example, monolayer MoS_2 loaded on an Si/SiO_2 surface has a band gap of 1.83 eV, which is suitable for low-power electronic devices. The high carrier mobility of MoS_2 devices of ~ 200 cm^2/V/s and high on/off current ratio of 10^{10} improves the performance of these devices two-fold over conventional field effect transistors. The thickness-dependent band gap property of MoS_2 layers was harnessed in photon-detecting devices. For example, bilayer MoS_2 (1.65 eV) is ideal for green light detection, whereas the trilayer (1.35 eV) is sensitive in the red region of the electromagnetic spectrum.[60] In addition, exfoliated MoS_2 is an interesting material for Li-ion batteries, because it shows a higher cycling ability than the bulk material (Table 4.3).[61] TMDs are also suitable for biomedical applications, as presented in Table 4.3.

The high surface-to-volume ratio of TMDs and their low cytotoxicity when compared to that of graphene[62] TMDs have highly beneficial characteristics for drug-delivery applications. The sensitive and strong NIR absorbance features of TMDs are also useful for imaging and photodynamic therapy for in vivo studies (Fig. 4.6).[63] Exfoliated MoS_2 modified with PEG loaded with an anticancer drug such as doxorubicin was used in combined photothermal and chemotherapy protocols.[64] In these modes, both heat and light energies are used to release/activate the drug. When MoS_2/PEG/doxorubicin was injected to the tumor site, tumor growth was inhibited over 20 days and tumor volume reduced to 15% when compared to the control, when laser irradiation at 808 nm was applied simultaneously (Fig. 4.6).[64] Similarly, nanolayers of WS_2 were used for image-guided photothermal treatments.[63] Activity of glucose oxidase or microbial activity was monitored by exploring the photoluminescence of MoS_2 nanosheets, which is sensitive to the intercalated ions (H^+, Li^+, Na^+, and K^+).[65] The above properties of TMDs provide a good arena for biologists because of their interactions with light (Table 4.3).

Table 4.3 Some Applications of MoS$_2$ Nanosheets

Material	Synthetic Route	Properties and Applications
MoS$_2$[66]	CVD	• Carrier mobility of ∼4 m^2/V/s • On/off ratio of ∼6 × 10^6 (FET)
MoS$_2$[67]	CVD	Monolayers of 120 μm in size
MoS$_2$[68]	Thermal decomposition of ammonium molybdate tetrahydrate	High mobility of 12.5 cm^2/V/s and an on/off current ratio of 10^5 in a transistor
Wafer scale-MoS$_2$[69]	Hydrothermal	• Hole mobility of ∼12.2 cm^2/V/s • On/off ratio of ∼10^3 in p-type transistor
Lignin/MoS$_2$[50]	Liquid-phase exfoliation	Li-ion battery cathode with capacity 110 mA.h/g
Thioglycolic acid (TGA)/MoS$_2$[51]	Liquid-phase exfoliation	Fluorescent, low cytotoxic MoS$_2$ quantum dots are used for bio-imaging
Butyl lithium/MoS$_2$[64]	Liquid-phase exfoliation followed by redispersion in water	Photodynamic therapy combined with chemotherapy
Chitosan/MoS$_2$[70]	Liquid-phase exfoliation after acid treatment with conc. H$_2$SO$_4$	Near-infrared (NIR) photodynamic therapy combined with chemotherapy
PEG/MoS$_2$[71]	Hydrothermal synthesis from ammonium molybdate tetrahydrate	Antibacterial agent in wound healing in combination with H$_2$O$_2$
NMP/MoS$_2$[72]	Liquid-phase exfoliation followed by gelation with peptides	NIR responsive gels (804 nm)

4.5 MXENES

Transition metal carbides, nitrides, and carbo-nitrides of few-atom-thick layers are called MXenes, because of their 2-D-layered structure, analogous to graphene. They are promising due to the combination of their high electrical conductivity and hydrophilic surface functions.[73] Rapid progress was made in this field soon after their discovery in 2011. MXenes are layers derived from MAX phases, where M is a transition metal (Ti, V, Cr, etc.), A is a group A element (e.g., Si, Al, etc., Groups 1, 2, 12, 13, 14, 15, 16, 17 in current IUPAC naming) and X represents carbon and/or nitrogen.[74] MAX phases have the general formula M$_{n+1}$AX$_n$, where n = 1, 2, or 3, e.g., Ti$_3$AlC$_2$. Nanolayers of MAX phases are obtained by acid etching out the "A" element (Al in Ti$_3$AlC$_2$), usually using HF, from the bulk material to yield M$_{n+1}$X$_n$ (see Fig. 1.9). These materials are then named as a graphene family member, MXenes (e.g., Ti$_3$C$_2$).[75] MXenes with M$_2$X and M$_4$X$_3$ composition are also known.[74]

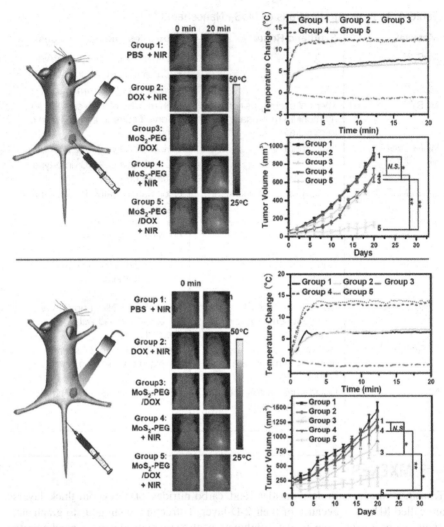

FIGURE 4.6

Photothermal therapy combined with chemotherapy was used to reduce tumor size in mice, using MoS$_2$ nanolayers as carriers as well as photoactive materials in combination with doxorubicin © John Wiley & Sons.[64]

4.5.1 PRODUCTION

In contrast to other 2-D materials discussed above, MXenes are produced by unconventional procedures. The MXene Ti$_3$C$_2$ will be used here as a model system to discuss these details. The precursor for Ti$_3$C$_2$ is Ti$_3$AlC$_2$, which is usually synthesized by dry ball milling of Ti$_2$AlC and TiC (1:1 ratio) for 24 hours,

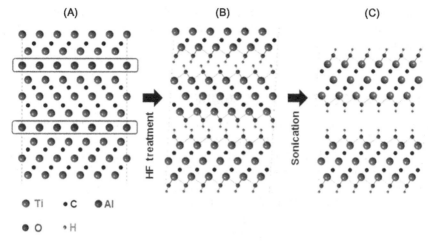

FIGURE 4.7

Etching of MAX phases using HF and subsequent delamination of MXenes by sonication ©John Wiley & Sons.[73]

followed by high-temperature heating under an inert atmosphere.[73] The resulting Ti_3AlC_2 powder is digested in HF to remove the Al layer and yield the MXene (Fig. 4.7; Eq. 4.3).[73] The ultra-thin layers of Ti_3C_2 are isolated by ultrasonication, and the layers have end groups of −OH or −F, according to Eqs. 4.4 and 4.5.[75]

$$Ti_3AlC_2 + 3HF \rightarrow AlF_3 + 3/2H_2 + Ti_3C_2 \qquad (4.3)$$

$$Ti_3C_2 + 2H_2O \rightarrow Ti_3C_2(OH)_2 + H_2 \qquad (4.4)$$

$$Ti_3C_2 + 2HF \rightarrow Ti_3C_2F_2 + H_2 \qquad (4.5)$$

Non-MAX phase precursors are also used to produce MXenes, as in the case of Mo_2C by etching Ga out of Mo_2Ga_2C by HF.[76] Exceptionally, synthesis of the zirconium carbide MXene ($Zr_3C_2T_x$; T = O, F, or OH) was accomplished by selective etching of Al_2C_3 (rather than just Al) from $Zr_3Al_3C_5$ (non-MAX phase) by HF.[77] In this process, the weakly bonded and easily hydrolyzable Al−C layer was removed from $Zr_3Al_3C_5$.[77] High-temperature etching with a fluoride salt was also reported to produce the MXene phase of Ti_4N_3, where Ti_4AlN_3 was heated with 59 wt% of potassium fluoride (KF), 29 wt% of lithium fluoride (LiF), and 12 wt% of sodium fluoride (NaF) under inert atmosphere at 550°C for 30 minutes.[78] The low stability of Ti−N bonds in the HF solution was avoided by this method.[78] Overall, most of these phases are synthesized by unique methods, depending on the availability of their precursors and the stability of their product to strong etching conditions. Computational studies are used as aids in most of these synthetic methods to find stable phases.[73] The MAX phase precursors are also synthesized from commercially available sources adding an additional step of synthesis, purification, and characterization to the overall process of MXene synthesis. At the end, sonication often is used to delaminate the material into the corresponding nanosheets.

4.5.2 CHARACTERIZATION TECHNIQUES

Since most of the MXenes have unique crystal structures and bonding patterns, discussion of characterization of each material is beyond the scope of this chapter. However, XRD, XPS, Raman spectroscopy, and electron microscopy are "must-be-done" characterization methods for MXenes as well. We focus on the characterization of Ti_3C_2, a popular member in the MXene family, in this section. The powder XRD pattern of Ti_3AlC_2 has well-defined crystalline patterns (Fig. 4.8A, bottom, *black* lines).[73] After etching with HF the pattern corresponding to amorphous powders appears around 24 Å (Fig. 4.8A, *second from the top* line). These patterns were compared with simulated XRD of $Ti_3C_2(OH)_2$ and $Ti_3C_2F_2$ (*third and forth from top* line).[73] Raman spectra of HF etched Ti_3AlC_2 (Fig. 4.8B) showed the disappearance of several peaks. XPS spectra (Fig. 4.8C) show the appearance of Ti−O (2p1) bonds, which confirms the formation of $Ti_3C_2(OH)_2$ units in the sample. Finally, SEM images (Fig. 4.8D) and TEM images show conclusive evidence for the layer morphology and expansion of the lattice after etching.[73]

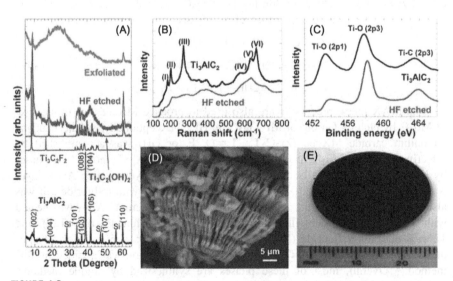

FIGURE 4.8

Characterization of TiC_2 MXene phase using XRD (A), Raman spectroscopy (B), XPS (C), and SEM (D). (E) Photograph of exfoliated TiC_2 MXene after cold pressing ©John Wiley & Sons.[73]

4.5.3 APPLICATIONS

MXenes have a wide range of applications in supercapacitors, electromagnetic interference shielding, batteries, and many more.[74] The facile intercalation of ions into MXenes was used to develop devices for electrochemical energy storage.[79] Titanium carbide showed electrochemical intercalation of various cations including Li^+, Na^+, K^+, NH_4^+, Mg^{2+}, and Al^{3+}. The capacitance of the optimized device was shown to be as high as 300 F/cm^3, outperforming the carbon-based capacitors.[79] It is also noted that $Ti_3C_2O_2$ is more efficient in storage of Li^+ when compared to its −F or −OH terminated counterparts, because of its likelihood of ion coordination.[80] In addition to the energy storage applications, MXenes are being tested for water purification,[81] biomedicine,[82,83] and electromagnetic shielding.[84]

Titanium carbide, an MXene with formula $(Ti_3C_2(OH/ONa)_xF_{2-x})$, selectively adsorbs lead ions, at concentrations below the World Health Organization's (WHO) regulations in drinking water (10 μg/L).[81] This characteristic was used to clean about 4500 kg water/kg of the MXene after several cycles of regeneration (using nitric acid).[81] Base treatment of Ti_3AlC_2 with tetramethyl ammonium hydroxide, instead of HF etching, was used to delaminate MAX phases. The resulting layers showed high absorption at the NIR region, which is promising for photothermal therapy applications.[82] In order to demonstrate the material's photothermic effect it was incubated with 4T1 murine breast cancer cells, where it showed no observable toxicity in the absence of light. However, when the cells were irradiated with an 808-nm laser, in the presence of an MXene that was modified with PEG for better dispersion in culture media, cell survivability was drastically reduced as a function of laser power, and at 1 W/cm^2 90% of the cells were killed within 5 minutes.[82] The antibacterial nature of $Ti_3C_2T_x$ (T = −OH, −F, or −O) was also reported and showed 98% cell death after 4-hour exposure (200 μg/mL), which was primarily due to cell wall disruption, mediated by oxidative stress.[83] The rich chemistry and unique structure of MXenes are promising for a variety of applications.

4.6 MONOATOMIC LAYERS

Inspired by the properties of single-layer carbon atoms in graphene, persistent attempts were made to make atomically thin layers of other elements. The most interesting member of this subfamily is phosphorene (also called 2-D-phosphane), because of its isolation from the naturally occurring allotrope of phosphorus (black phosphorus; Fig. 4.9).[85,86] Borophene is the single-atom layer of boron, with metallic characteristics, synthesized so far only on a silver surface.[87] Likewise, single-atom layers of germanium (germancene) are also synthesized on Pt surface using molecular beam epitaxy.[88] Most of these members were predicted to have applications in the semiconductor field, but continuous efforts are

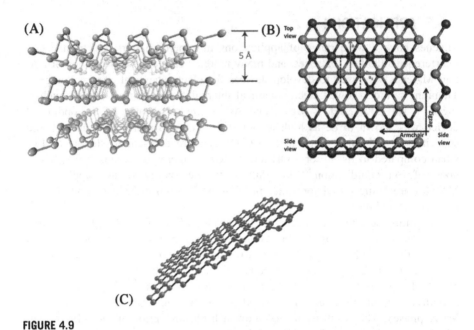

FIGURE 4.9

Structures of (A) phosphorene (©Nature Publishing Group),[90] (B) borophene (©Royal Chemical Society),[91] and (C) silicene (©Nature Publishing Group).[92]

required to prove their stability in free form and the possibility of scalable production. Solution-phase synthesis of silicene is an example of such progress, where phenyl magnesium bromide was used to exfoliate silicon crystals.[89]

4.6.1 SYNTHESIS OR EXFOLIATION

Most of the nongraphene single-atom layers are not fully synthesized in free-standing form except phosphorene. The possibility of exfoliation of black phosphorus in solvents or by the Scotch tape method makes it akin to graphene.[93,94] Black phosphorus was exfoliated in dry-NMP by ultrasonication to yield few-layer (3–5) black phosphorus with lateral size of ~ 100 nm.[93] However, phosphorene is highly sensitive to air and water, and exposure to ambient air results in its degradation, which is a major disadvantage.[95] N-cyclohexyl-2-pyrrolidone was proposed as a strongly protecting medium for exfoliated black phosphorus over NMP or other solvents. The solvation shell of N-cyclohexyl-2-pyrrolidone has a protecting effect on phosphorene and only 10% degradation was observed in this system after exposure to ambient conditions for 400 hours, but the stability is very close to black phosphorus nanolayers in NMP with $\sim 80\%$ retention. In contrast, exfoliation of the black phosphorus was achieved in deoxygenated surfactant water systems (2% SDS).[96] Thus, oxygen, rather than water alone is responsible for the degradation of phosphorene, hence, exposure to ambient conditions requires further protection.[96] Exfoliation of the other monoatomic layers is not yet known.

4.6.2 CHARACTERIZATION

Phosphorene was characterized primarily by Raman spectroscopy and microscopy techniques. The Raman spectrum of black phosphorus after exfoliation is different from that of the bulk, especially below two layers.[97] Black phosphorus possesses three Raman active modes of A_g^1 (~ 360 cm^{-1}), B_{2g} (~ 445 cm^{-1}), and A_g^2 (~ 465 cm^{-1}). Exfoliation to monolayer phosphorene results in a slight shift of A_g^2 (~ 465 cm^{-1}) to higher wavenumbers (~ 1 cm^{-1} change).[97] The two-layer phosphorene shows broadening of the A_g^2 bands, when compared to the bulk, but there are no changes in the spectra of the trilayer or higher-layer samples. Microscopy studies were carried out to confirm the exfoliation and oxidative damage to the sheets. Exfoliation of other monoatomic layers was not obtainable in freestanding form, and thus only high-resolution imaging, such as scanning tunneling microscopy or scanning probe microscopy images, are available now (Fig. 4.10).[87,88]

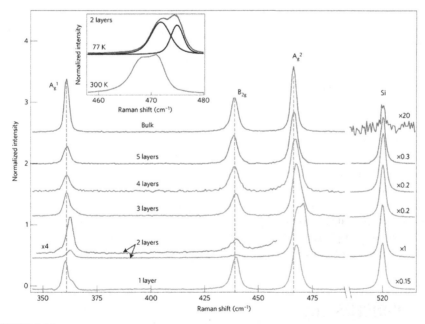

FIGURE 4.10

Raman spectra of black phosphorus at different thickness (number of layers) © Nature Publishing Group.[97]

Table 4.4 Some Applications of Black Phosphorus Nanosheets

Material	Synthetic Route	Properties and Applications
Few-layer phosphorene[90]	Micromechanical exfoliation using Scotch tape	• Hole mobility of $\sim$1000 cm^2/V/s • On/off ratio of $\sim$10^5 in a transistor
Few-layer phosphorene[100]	Dry transfer	• On/off ratio of $\sim$10^3 in p-type transistor • Visible to NIR (940 nm) detection range
Few-layer phosphorene[101]	Freshly cleaved black phosphorus	• Thermo-responsive FET with 5–100 μV/K at 77 K sensitivity (Seebeck coefficient)
Few-layer phosphorene[102]	Micromechanical exfoliation using Scotch tape	• FET with $\sim$10^3 on/off ratio • Carrier mobility of $\sim$100 cm^2/V/s • Operated in ambient conditions for 2 weeks
Few-layer phosphorene[103]	Micromechanical exfoliation using Scotch tape	• FET NO$_2$ sensor with 20 ppb detection limit
Few-layer phosphorene[104]	Dry ball milling	• Composites with graphite • Li-ion battery materials with 600 mA.h/g capacity • Stable (100%) for >100 cycles
Few-layer phosphorene/ graphene hybrid[105]	Liquid-phase exfoliation in NMP	• Na-ion battery with 2440 mA.h/g current capacity • 83% retention after 100 cycles

4.6.3 APPLICATIONS

Isolation of these single-atom-layered materials is often motivated by their promising electronic applications.[98] Black phosphorus in monolayer form has a direct band gap of $\sim$1.0 eV, in comparison to the bulk form with a value of $\sim$0.3 eV.[99] Unlike other 2-D materials, the band gap does not vary with the thickness of phosphorene, which makes the exfoliated few-layer phosphorene a promising material in transistors, inverters, solar cells, and for the manufacturing of flexible electronics (Table 4.4). These properties were harnessed in the construction of field effect transistors and photodetectors, in combination with other 2-D materials.[86] However, protection of phosphorene from degradation by oxygen still needs to be addressed. Selected applications of few-layer phosphorene are listed in Table 4.4.

4.7 CONCLUSIONS

Inorganic analogues of graphene with rich chemistry and promising applications are examined in this chapter at an introductory level. Materials with commercially and readily available precursors are often studied thoroughly. Liquid-phase exfoliation of those materials yields solution-processable nanolayers which may find applications beyond electronics.[81−83] Unlike graphene, the inorganic layers are more reactive and thus sensitive to the environment of exfoliation or fabrication. For example, *h*-BN nanolayers are prone to hydrolysis under ultrasonication conditions,[16] and black phosphorus nanolayers are sensitive to ambient oxygen.[94] Challenges to be addressed include these issues, as well as the scalable production of the new classes of inorganic layered materials, preferably in aqueous media or other low-temperature boiling solvents. Theoretical studies of how these materials interact with small molecules, ions, polymers, proteins, and nucleic acids would be very useful and could readily guide experimental studies in promising directions.

REFERENCES

1. Novoselov, K. S.; Jiang, D.; Schedin, F.; Booth, T. J.; Khotkevich, V. V.; Morozov, S. V.; Geim, A. K. Two-Dimensional Atomic Crystals. *Proc. Natl. Acad. Sci. USA* **2005,** *102* (30), 10451−10453.
2. Nicolosi, V.; Chhowalla, M.; Kanatzidis, M. G.; Strano, M. S.; Coleman, J. N. Liquid Exfoliation of Layered Materials. *Science* **2013,** *340* (6139), 1226419.
3. Wang, X.; Pakdel, A.; Zhang, J.; Weng, Q.; Zhai, T.; Zhi, C.; Golberg, D.; Bando, Y. Large-Surface-Area BN Nanosheets and Their Utilization in Polymeric Composites with Improved Thermal and Dielectric Properties. *Nanoscale Res. Lett.* **2012,** *7* (1), 662.
4. Lipp, A.; Schwetz, K. A.; Hunold, K. Hexagonal Boron Nitride: Fabrication, Properties and Applications. *J. Eur. Ceram. Soc.* **1989,** *5* (1), 3−9.
5. Ertug, B. Powder Preparation, Properties and Industrial Applications of Hexagonal Boron Nitride. In *Sintering Applications;* Ertuğ, B., Ed.; InTech: Rijeka, 2013; p Ch. 02.
6. Pease, R. S. An X-Ray Study of Boron Nitride. *Acta. Crystallogr.* **1952,** *5,* 356−361.
7. Xu, M.; Fujita, D.; Chen, H.; Hanagata, N. Formation of Monolayer and Few-Layer Hexagonal Boron Nitride Nanosheets via Surface Segregation. *Nanoscale* **2011,** *3* (7), 2854−2858.
8. Kim, K. K.; Hsu, A.; Jia, X.; Kim, S. M.; Shi, Y.; Hofmann, M.; Nezich, D.; Rodriguez-Nieva, J. F.; Dresselhaus, M.; Palacios, T. Synthesis of Monolayer Hexagonal Boron Nitride on Cu Foil Using Chemical Vapor Deposition. *Nano. Lett.* **2011,** *12* (1), 161−166.
9. Lin, Y.; Connell, J. W. Advances in 2D Boron Nitride Nanostructures: Nanosheets, Nanoribbons, Nanomeshes, and Hybrids with Graphene. *Nanoscale* **2012,** *4* (22), 6908−6939.
10. Mukhopadhyay, T. K.; Datta, A. Deciphering the Role of Solvents in the Liquid Phase Exfoliation of Hexagonal Boron Nitride: A Molecular Dynamics Simulation Study. *J. Phys. Chem. C* **2016,** *121* (18), 10210−10223.
11. Zhou, K.-G.; Mao, N.-N.; Wang, H.-X.; Peng, Y.; Zhang, H.-L.; Mixed-Solvent, A. Strategy for Efficient Exfoliation of Inorganic Graphene Analogues. *Angew. Chem. Int. Ed.* **2011,** *50* (46), 10839−10842.
12. Hernandez, Y.; Nicolosi, V.; Lotya, M.; Blighe, F. M.; Sun, Z.; De, S.; McGovern, I. T.; Holland, B.; Byrne, M.; Gun'Ko, Y. K., et al. High-yield production of graphene by liquid-phase exfoliation of graphite. *Nat. Nano* **2008,** *3* (9), 563−568.

13. Chen, X.; Dobson, J. F.; Raston, C. L. Vortex Fluidic Exfoliation of Graphite and Boron Nitride. *Chem. Commun.* **2012,** *48* (31), 3703–3705.

14. Li, L. H.; Glushenkov, A. M.; Hait, S. K.; Hodgson, P.; Chen, Y. High-Efficient Production of Boron Nitride Nanosheets via an Optimized Ball Milling Process for Lubrication in Oil. *Sci. Rep.* **2014,** *4*, 7288.

15. Pakdel, A.; Zhi, C.; Bando, Y.; Nakayama, T.; Golberg, D. Boron Nitride Nanosheet Coatings with Controllable Water Repellency. *ACS Nano* **2011,** *5* (8), 6507–6515.

16. Lin, Y.; Williams, T. V.; Xu, T.-B.; Cao, W.; Elsayed-Ali, H. E.; Connell, J. W. Aqueous Dispersions of Few-Layered and Monolayered Hexagonal Boron Nitride Nanosheets from Sonication-Assisted Hydrolysis: Critical Role of Water. *J. Phys. Chem. C* **2011,** *115* (6), 2679–2685.

17. Yoshizaki, T.; Watanabe, H.; Nakagawa, T. *Kinetics and Mechanisms of Hydrolysis of Borazine Derivatives in Aqueous Dioxane;* Shionogi and Co., Ltd.: Osaka, 1968.

18. Lu, F.; Wang, F.; Gao, W.; Huang, X.; Zhang, X.; Li, Y. Aqueous Soluble Boron Nitride Nanosheets via Anionic Compound-Assisted Exfoliation. *Mater. Exp.* **2013,** *3* (2), 144–150.

19. Bari, R.; Parviz, D.; Khabaz, F.; Klaassen, C. D.; Metzler, S. D.; Hansen, M. J.; Khare, R.; Green, M. J. Liquid Phase Exfoliation and Crumpling of Inorganic Nanosheets. *Phys. Chem. Chem. Phys.* **2015,** *17* (14), 9383–9393.

20. Rafiei-Sarmazdeh, Z.; Jafari, S. H.; Ahmadi, S. J.; Zahedi-Dizaji, S. M. Large-Scale Exfoliation of Hexagonal Boron Nitride with Combined Fast Quenching and Liquid Exfoliation Strategies. *J. Mater. Sci.* **2016,** *51* (6), 3162–3169.

21. Ge, Y.; Wang, J.; Shi, Z.; Yin, J. Gelatin-Assisted Fabrication of Water-Dispersible Graphene and Its Inorganic Analogues. *J. Mater. Chem.* **2012,** *22* (34), 17619–17624.

22. Lin, Y.; Williams, T. V.; Connell, J. W. Soluble, Exfoliated Hexagonal Boron Nitride Nanosheets. *J. Phys. Chem. Lett.* **2010,** *1* (1), 277–283.

23. Song, L.; Ci, L.; Lu, H.; Sorokin, P. B.; Jin, C.; Ni, J.; Kvashnin, A. G.; Kvashnin, D. G.; Lou, J.; Yakobson, B. I. Large Scale Growth and Characterization of Atomic Hexagonal Boron Nitride Layers. *Nano Lett.* **2010,** *10* (8), 3209–3215.

24. Gorbachev, R. V.; Riaz, I.; Nair, R. R.; Jalil, R.; Britnell, L.; Belle, B. D.; Hill, E. W.; Novoselov, K. S.; Watanabe, K.; Taniguchi, T. Hunting for Monolayer Boron Nitride: Optical and Raman signatures. *Small* **2011,** *7* (4), 465–468.

25. Song, W.-L.; Wang, P.; Cao, L.; Anderson, A.; Meziani, M. J.; Farr, A. J.; Sun, Y.-P. Polymer/Boron Nitride Nanocomposite Materials for Superior Thermal Transport Performance. *Angew. Chem. Int. Ed.* **2012,** *51* (26), 6498–6501.

26. Yu, J.; Huang, X.; Wu, C.; Wu, X.; Wang, G.; Jiang, P. Interfacial Modification of Boron Nitride Nanoplatelets for Epoxy Composites with Improved Thermal Properties. *Polymer. (Guildf).* **2012,** *53* (2), 471–480.

27. Kiran, M. S. R. N.; Raidongia, K.; Ramamurty, U.; Rao, C. N. R. Improved Mechanical Properties of Polymer Nanocomposites Incorporating Graphene-Like BN: Dependence on the Number of BN Layers. *Scripta Mater.* **2011,** *64* (6), 592–595.

28. Dean, C. R.; Young, A. F.; Meric, I.; Lee, C.; Wang, L.; Sorgenfrei, S.; Watanabe, K.; Taniguchi, T.; Kim, P.; Shepard, K. L.; Hone, J. Boron Nitride Substrates for High-Quality Graphene Electronics. *Nat. Nano* **2010,** *5* (10), 722–726.

29. Yankowitz, M.; Xue, J.; Cormode, D.; Sanchez-Yamagishi, J. D.; Watanabe, K.; Taniguchi, T.; Jarillo-Herrero, P.; Jacquod, P.; LeRoy, B. J. Emergence of

Superlattice Dirac Points in Graphene on Hexagonal Boron Nitride. *Nat. Phys.* **2012,** *8* (5), 382–386.

30. Wang, L.; Sun, C.; Xu, L.; Qian, Y. Convenient Synthesis and Applications of Gram Scale Boron Nitride Nanosheets. *Cat. Sci. Technol.* **2011,** *1* (7), 1119–1123.

31. Lin, Y.; Bunker, C. E.; Fernando, K. A. S.; Connell, J. W. Aqueously Dispersed Silver Nanoparticle-Decorated Boron Nitride Nanosheets for Reusable, Thermal Oxidation-Resistant Surface Enhanced Raman Spectroscopy (SERS) Devices. *ACS Appl. Mater. Interf.* **2012,** *4* (2), 1110–1117.

32. Chen, Y.; Tan, C.; Zhang, H.; Wang, L. Two-Dimensional Graphene Analogues for Biomedical Applications. *Chem. Soc. Rev.* **2015,** *44* (9), 2681–2701.

33. Zhi, C.; Bando, Y.; Tang, C.; Kuwahara, H.; Golberg, D. Large-scale Fabrication of Boron Nitride Nanosheets and Their Utilization in Polymeric Composites with Improved Thermal and Mechanical Properties. *Adv. Mater.* **2009,** *21* (28), 2889–2893.

34. Weng, Q.; Wang, B.; Wang, X.; Hanagata, N.; Li, X.; Liu, D.; Wang, X.; Jiang, X.; Bando, Y.; Golberg, D. Highly Water-Soluble, Porous, and Biocompatible Boron Nitrides for Anticancer Drug Delivery. *ACS Nano* **2014,** *8* (6), 6123–6130.

35. White, R. M.; Lucovsky, G. Chemical Bonding and Structure in Layered Transition Metal Dichalcogenides. *Solid State Commun.* **1972,** *11* (10), 1369–1373.

36. Mak, K. F.; Lee, C.; Hone, J.; Shan, J.; Heinz, T. F. Atomically Thin MoS_2: A New Direct-Gap Semiconductor. *Phys. Rev. Lett.* **2010,** *105* (13), 136805.

37. Rao, C.; Ramakrishna Matte, H.; Maitra, U. Graphene Analogues of Inorganic Layered Materials. *Angew. Chem. Int. Ed.* **2013,** *52* (50), 13162–13185.

38. Benameur, M.; Radisavljevic, B.; Heron, J.; Sahoo, S.; Berger, H.; Kis, A. Visibility of Dichalcogenide Nanolayers. *Nanotechnology* **2011,** *22* (12), 125706.

39. Jawaid, A.; Nepal, D.; Park, K.; Jespersen, M.; Qualley, A.; Mirau, P.; Drummy, L. F.; Vaia, R. A. Mechanism for Liquid Phase Exfoliation of MoS_2. *Chem. Mater.* **2016,** *28* (1), 337–348.

40. Gupta, A.; Arunachalam, V.; Vasudevan, S. Liquid-Phase Exfoliation of MoS_2 Nanosheets: The Critical Role of Trace Water. *J. Phys. Chem. Lett.* **2016,** *7* (23), 4884–4890.

41. Varrla, E.; Backes, C.; Paton, K. R.; Harvey, A.; Gholamvand, Z.; McCauley, J.; Coleman, J. N. Large-Scale Production of Size-Controlled MoS_2 Nanosheets by Shear Exfoliation. *Chem. Mater.* **2015,** *27* (3), 1129–1139.

42. Yao, Y.; Tolentino, L.; Yang, Z.; Song, X.; Zhang, W.; Chen, Y.; Wong, C.-P. High-Concentration Aqueous Dispersions of MoS_2. *Adv. Funct. Mater.* **2013,** *23* (28), 3577–3583.

43. Smith, R. J.; King, P. J.; Lotya, M.; Wirtz, C.; Khan, U.; De, S.; O'Neill, A.; Duesberg, G. S.; Grunlan, J. C.; Moriarty, G., et al. Large-Scale Exfoliation of Inorganic Layered Compounds in Aqueous Surfactant Solutions. *Adv. Mater.* **2011,** *23* (34), 3944–3948.

44. Forsberg, V.; Zhang, R.; Bäckström, J.; Dahlström, C.; Andres, B.; Norgren, M.; Andersson, M.; Hummelgård, M.; Olin, H. Exfoliated MoS_2 in Water Without Additives. *PLoS One* **2016,** *11* (4), e0154522.

45. Kim, J.; Kwon, S.; Cho, D.-H.; Kang, B.; Kwon, H.; Kim, Y.; Park, S. O.; Jung, G. Y.; Shin, E.; Kim, W.-G. Direct Exfoliation and Dispersion of Two-Dimensional Materials in Pure Water via Temperature Control. *Nat. Commun.* **2015,** *6*, 8294.

46. Pattammattel, A.; Kumar, C. V. Kitchen Chemistry 101: Multigram Production of High Quality Biographene in a Blender with Edible Proteins. *Adv. Funct. Mater.* **2015,** *25* (45), 7088−7098.

47. Yao, Y.; Lin, Z.; Li, Z.; Song, X.; Moon, K.-S.; Wong, C.-P. Large-Scale Production of Two-Dimensional Nanosheets. *J. Mater. Chem.* **2012,** *22* (27), 13494−13499.

48. Guardia, L.; Paredes, J. I.; Rozada, R.; Villar-Rodil, S.; Martinez-Alonso, A.; Tascon, J. M. D. Production of Aqueous Dispersions of Inorganic Graphene Analogues by Exfoliation and Stabilization with Non-Ionic Surfactants. *RSC Adv.* **2014,** *4* (27), 14115−14127.

49. Liu, N.; Kim, P.; Kim, J. H.; Ye, J. H.; Kim, S.; Lee, C. J. Large-Area Atomically Thin MoS_2 Nanosheets Prepared Using Electrochemical Exfoliation. *ACS Nano* **2014,** *8* (7), 6902−6910.

50. Liu, W.; Zhao, C.; Zhou, R.; Zhou, D.; Liu, Z.; Lu, X. Lignin-Assisted Exfoliation of Molybdenum Disulfide in Aqueous Media and Its Application in Lithium Ion Batteries. *Nanoscale* **2015,** *7* (21), 9919−9926.

51. Anbazhagan, R.; Wang, H.-J.; Tsai, H.-C.; Jeng, R.-J. Highly Concentrated MoS2 Nanosheets in Water Achieved by Thioglycolic Acid as Stabilizer and Used as Biomarkers. *RSC Adv.* **2014,** *4* (81), 42936−42941.

52. Guan, G.; Zhang, S.; Liu, S.; Cai, Y.; Low, M.; Teng, C. P.; Phang, I. Y.; Cheng, Y.; Duei, K. L.; Srinivasan, B. M. Protein Induces Layer-By-Layer Exfoliation of Transition Metal Dichalcogenides. *J. Am. Chem. Soc.* **2015,** *137* (19), 6152−6155.

53. Sim, H.; Lee, J.; Park, B.; Kim, S. J.; Kang, S.; Ryu, W.; Jun, S. C. High-Concentration Dispersions of Exfoliated MoS_2 Sheets Stabilized by Freeze-Dried Silk Fibroin Powder. *Nano Res.* **2016,** *9* (6), 1709−1722.

54. Xiao, X.; Song, H.; Lin, S.; Zhou, Y.; Zhan, X.; Hu, Z.; Zhang, Q.; Sun, J.; Yang, B., et al. Scalable Salt-Templated Synthesis of Two-Dimensional Transition Metal Oxides. *Nat. Commun.* **2016,** *7*, 11296.

55. Shen, J.; He, Y.; Wu, J.; Gao, C.; Keyshar, K.; Zhang, X.; Yang, Y.; Ye, M.; Vajtai, R.; Lou, J. Liquid Phase Exfoliation of Two-Dimensional Materials by Directly Probing and Matching Surface Tension Components. *Nano Lett.* **2015,** *15* (8), 5449−5454.

56. Backes, C.; Smith, R. J.; McEvoy, N.; Berner, N. C.; McCloskey, D.; Nerl, H. C.; O'Neill, A.; King, P. J.; Higgins, T.; Hanlon, D. Edge and Confinement Effects Allow In Situ Measurement of Size and Thickness of Liquid-Exfoliated Nanosheets. *Nat. Commun.* **2014,** *5*, 4576.

57. Li, H.; Zhang, Q.; Yap, C. C. R.; Tay, B. K.; Edwin, T. H. T.; Olivier, A.; Baillargeat, D. From Bulk to Monolayer MoS_2: Evolution of Raman Scattering. *Adv. Funct. Mater.* **2012,** *22* (7), 1385−1390.

58. Boukhicha, M.; Calandra, M.; Measson, M.-A.; Lancry, O.; Shukla, A. Anharmonic Phonons in Few-Layer MoS_2: Raman Spectroscopy of Ultralow Energy Compression and Shear Modes. *Phys. Rev. B* **2013,** *87* (19), 195316.

59. Gordon, R.; Yang, D.; Crozier, E.; Jiang, D.; Frindt, R. Structures of Exfoliated Single Layers of WS_2, MoS_2, and $MoSe_2$ in Aqueous Suspension. *Phys. Rev. B* **2002,** *65* (12), 125407.

60. Yuanfang, Y.; Feng, M.; Jun, H.; Zhenhua, N. Photodetecting and Light-Emitting Devices Based on Two-Dimensional Materials. *Chin. Phys. B* **2017,** *26* (3), 036801.

61. Feng, C.; Ma, J.; Li, H.; Zeng, R.; Guo, Z.; Liu, H. Synthesis of Molybdenum Disulfide (MoS_2) for Lithium Ion Battery Applications. *Mater. Res. Bull.* **2009,** *44* (9), 1811−1815.

62. Teo, W. Z.; Chng, E. L. K.; Sofer, Z.; Pumera, M. Cytotoxicity of Exfoliated Transition-Metal Dichalcogenides (MoS_2, WS_2, and WSe_2) Is Lower Than That of Graphene and its Analogues. *Chem. A Eur. J.* **2014,** *20* (31), 9627−9632.

63. Chen, Y.; Tan, C.; Zhang, H.; Wang, L. Two-Dimensional Graphene Analogues for Biomedical Applications. *Chem. Soc. Rev.* **2015,** *44* (9), 2681−2701.

64. Liu, T.; Wang, C.; Gu, X.; Gong, H.; Cheng, L.; Shi, X.; Feng, L.; Sun, B.; Liu, Z. Drug Delivery with PEGylated MoS_2 Nano-sheets for Combined Photothermal and Chemotherapy of Cancer. *Adv. Mater.* **2014,** *26* (21), 3433−3440.

65. Ou, J. Z.; Chrimes, A. F.; Wang, Y.; Tang, S.-y; Strano, M. S.; Kalantar-zadeh, K. Ion-Driven Photoluminescence Modulation of Quasi-Two-Dimensional MoS_2 Nanoflakes for Applications in Biological Systems. *Nano Lett.* **2014,** *14* (2), 857−863.

66. Ling, X.; Lee, Y.-H.; Lin, Y.; Fang, W.; Yu, L.; Dresselhaus, M. S.; Kong, J. Role of the Seeding Promoter in MoS_2 Growth by Chemical Vapor Deposition. *Nano Lett.* **2014,** *14* (2), 464−472.

67. Lin, Y.-C.; Lu, N.; Perea-Lopez, N.; Li, J.; Lin, Z.; Peng, X.; Lee, C. H.; Sun, C.; Calderin, L.; Browning, P. N.; Bresnehan, M. S.; Kim, M. J.; Mayer, T. S.; Terrones, M.; Robinson, J. A. Direct Synthesis of van der Waals Solids. *ACS Nano* **2014,** *8* (4), 3715−3723.

68. Pu, J.; Yomogida, Y.; Liu, K.-K.; Li, L.-J.; Iwasa, Y.; Takenobu, T. Highly Flexible MoS_2 Thin-Film Transistors With Ion Gel Dielectrics. *Nano Lett.* **2012,** *12* (8), 4013−4017.

69. Tao, J.; Chai, J.; Lu, X.; Wong, L. M.; Wong, T. I.; Pan, J.; Xiong, Q.; Chi, D.; Wang, S. Growth of Wafer-Scale MoS_2 Monolayer by Magnetron Sputtering. *Nanoscale* **2015,** *7* (6), 2497−2503.

70. Yin, W.; Yan, L.; Yu, J.; Tian, G.; Zhou, L.; Zheng, X.; Zhang, X.; Yong, Y.; Li, J.; Gu, Z.; Zhao, Y. High-Throughput Synthesis of Single-Layer MoS_2 Nanosheets as a Near-Infrared Photothermal-Triggered Drug Delivery for Effective Cancer Therapy. *ACS Nano* **2014,** *8* (7), 6922−6933.

71. Yin, W.; Yu, J.; Lv, F.; Yan, L.; Zheng, L. R.; Gu, Z.; Zhao, Y. Functionalized Nano-MoS_2 with Peroxidase Catalytic and Near-Infrared Photothermal Activities for Safe and Synergetic Wound Antibacterial Applications. *ACS Nano* **2016,** *10* (12), 11000−11011.

72. Singh, A.; Kapil, N.; Yenuganti, M.; Das, D. Exfoliated Sheets of MoS_2 Trigger Formation of Aqueous Gels with Acute NIR Light Responsiveness. *Chem. Commun.* **2016,** *52* (97), 14043−14046.

73. Naguib, M.; Kurtoglu, M.; Presser, V.; Lu, J.; Niu, J.; Heon, M.; Hultman, L.; Gogotsi, Y.; Barsoum, M. W. Two-Dimensional Nanocrystals Produced by Exfoliation of Ti_3AlC_2. *Adv. Mater.* **2011,** *23* (37), 4248−4253.

74. Anasori, B.; Lukatskaya, M. R.; Gogotsi, Y. 2D Metal Carbides and Nitrides (MXenes) for Energy Storage. *Nat. Rev. Mater.* **2017,** *2*, 16098.

75. Mashtalir, O.; Naguib, M.; Dyatkin, B.; Gogotsi, Y.; Barsoum, M. W. Kinetics of Aluminum Extraction From Ti_3AlC_2 in Hydrofluoric Acid. *Mater. Chem. Phys.* **2013,** *139* (1), 147−152.

76. Halim, J.; Kota, S.; Lukatskaya, M. R.; Naguib, M.; Zhao, M.-Q.; Moon, E. J.; Pitock, J.; Nanda, J.; May, S. J.; Gogotsi, Y., et al. Synthesis and Characterization of 2D Molybdenum Carbide (MXene). *Adv. Funct. Mater.* **2016,** *26* (18), 3118–3127.

77. Zhou, J.; Zha, X.; Chen, F. Y.; Ye, Q.; Eklund, P.; Du, S.; Huang, Q. A Two-Dimensional Zirconium Carbide by Selective Etching of Al3C3 from Nanolaminated Zr$_3$Al$_3$C$_5$. *Angew. Chem.* **2016,** *128* (16), 5092–5097.

78. Urbankowski, P.; Anasori, B.; Makaryan, T.; Er, D.; Kota, S.; Walsh, P. L.; Zhao, M.; Shenoy, V. B.; Barsoum, M. W.; Gogotsi, Y. Synthesis of Two-Dimensional Titanium Nitride Ti$_4$N$_3$ (MXene). *Nanoscale* **2016,** *8* (22), 11385–11391.

79. Lukatskaya, M. R.; Mashtalir, O.; Ren, C. E.; Dall'Agnese, Y.; Rozier, P.; Taberna, P. L.; Naguib, M.; Simon, P.; Barsoum, M. W.; Gogotsi, Y. Cation Intercalation and High Volumetric Capacitance of Two-Dimensional Titanium Carbide. *Science* **2013,** *341* (6153), 1502–1505.

80. Xie, Y.; Naguib, M.; Mochalin, V. N.; Barsoum, M. W.; Gogotsi, Y.; Yu, X.; Nam, K.-W.; Yang, X.-Q.; Kolesnikov, A. I.; Kent, P. R. C. Role of Surface Structure on Li-Ion Energy Storage Capacity of Two-Dimensional Transition-Metal Carbides. *J. Am. Chem. Soc.* **2014,** *136* (17), 6385–6394.

81. Peng, Q.; Guo, J.; Zhang, Q.; Xiang, J.; Liu, B.; Zhou, A.; Liu, R.; Tian, Y. Unique Lead Adsorption Behavior of Activated Hydroxyl Group in Two-Dimensional Titanium Carbide. *J. Am. Chem. Soc.* **2014,** *136* (11), 4113–4116.

82. Xuan, J.; Wang, Z.; Chen, Y.; Liang, D.; Cheng, L.; Yang, X.; Liu, Z.; Ma, R.; Sasaki, T.; Geng, F. Organic-Base-Driven Intercalation and Delamination for the Production of Functionalized Titanium Carbide Nanosheets with Superior Photothermal Therapeutic Performance. *Angew. Chem.* **2016,** *128* (47), 14789–14794.

83. Rasool, K.; Helal, M.; Ali, A.; Ren, C. E.; Gogotsi, Y.; Mahmoud, K. A. Antibacterial Activity of Ti$_3$C$_2$T$_x$ MXene. *ACS Nano* **2016,** *10* (3), 3674–3684.

84. Shahzad, F.; Alhabeb, M.; Hatter, C. B.; Anasori, B.; Hong, S. M.; Koo, C. M.; Gogotsi, Y. Electromagnetic Interference Shielding with 2D Transition Metal Carbides (MXenes). *Science* **2016,** *353* (6304), 1137–1140.

85. Eswaraiah, V.; Zeng, Q.; Long, Y.; Liu, Z. Black Phosphorus Nanosheets: Synthesis, Characterization and Applications. *Small* **2016,** *12* (26), 3480–3502.

86. Carvalho, A.; Wang, M.; Zhu, X.; Rodin, A. S.; Su, H.; Neto, A. H. C. Phosphorene: From Theory to Applications. *Nat. Rev. Mater.* **2016,** *1*, 16061.

87. Mannix, A. J.; Zhou, X.-F.; Kiraly, B.; Wood, J. D.; Alducin, D.; Myers, B. D.; Liu, X.; Fisher, B. L.; Santiago, U.; Guest, J. R. Synthesis of Borophenes: Anisotropic, Two-Dimensional Boron Polymorphs. *Science* **2015,** *350* (6267), 1513–1516.

88. Dávila, M. E.; Xian, L.; Cahangirov, S.; Rubio, A.; Lay, G. L. Germanene: A Novel Two-Dimensional Germanium Allotrope Akin to Graphene and Silicene. *New. J. Phys.* **2014,** *16* (9), 095002.

89. Sugiyama, Y.; Okamoto, H.; Mitsuoka, T.; Morikawa, T.; Nakanishi, K.; Ohta, T.; Nakano, H. Synthesis and Optical Properties of Monolayer Organosilicon Nanosheets. *J. Am. Chem. Soc.* **2010,** *132* (17), 5946–5947.

90. Li, L.; Yu, Y.; Ye, G. J.; Ge, Q.; Ou, X.; Wu, H.; Feng, D.; Chen, X. H.; Zhang, Y. Black Phosphorus Field-Effect Transistors. *Nat. Nano* **2014,** *9* (5), 372–377.

91. Sun, H.; Li, Q.; Wan, X. G. First-Principles Study of Thermal Properties of Borophene. *Phys. Chem. Chem. Phys.* **2016,** *18* (22), 14927–14932.

92. Tao, L.; Cinquanta, E.; Chiappe, D.; Grazianetti, C.; Fanciulli, M.; Dubey, M.; Molle, A.; Akinwande, D. Silicene Field-Effect Transistors Operating at Room Temperature. *Nat. Nano* **2015,** *10* (3), 227–231.

93. Brent, J. R.; Savjani, N.; Lewis, E. A.; Haigh, S. J.; Lewis, D. J.; O'Brien, P. Production of Few-Layer Phosphorene by Liquid Exfoliation of Black Phosphorus. *Chem. Commun.* **2014,** *50* (87), 13338–13341.

94. Liu, H.; Neal, A. T.; Zhu, Z.; Xu, X.; Tomanek, D.; Ye, P. D.; Luo, Z. Phosphorene: An Unexplored 2D Semiconductor with a High Hole Mobility. *ACS Nano* **2014,** *8* (4), 4033–4041.

95. Hanlon, D.; Backes, C.; Doherty, E.; Cucinotta, C.S.; Berner, N.C.; Boland, C.; Lee, K.; Lynch, P.; Gholamvand, Z.; Harvey, A. Liquid exfoliation of solvent-stabilised black phosphorus: applications beyond electronics. *arXiv preprint arXiv:1501.01881* **2015**.

96. Kang, J.; Wells, S. A.; Wood, J. D.; Lee, J.-H.; Liu, X.; Ryder, C. R.; Zhu, J.; Guest, J. R.; Husko, C. A.; Hersam, M. C. Stable Aqueous Dispersions of Optically and Electronically Active Phosphorene. *Proc. Natl. Acad. Sci.* **2016,** *113* (42), 11688–11693.

97. Favron, A.; Gaufrès, E.; Fossard, F.; Phaneuf-L'Heureux, A.-L.; Tang, N. Y.; Lévesque, P. L.; Loiseau, A.; Leonelli, R.; Francoeur, S.; Martel, R. Photooxidation and Quantum Confinement Effects in Exfoliated Black Phosphorus. *Nat. Mater.* **2015,** *14* (8), 826–832.

98. Eswaraiah, V.; Zeng, Q.; Long, Y.; Liu, Z. Black Phosphorus Nanosheets: Synthesis, Characterization and Applications. *Small* **2016,** *12* (26), 3480–3502.

99. Kim, J.; Baik, S. S.; Ryu, S. H.; Sohn, Y.; Park, S.; Park, B.-G.; Denlinger, J.; Yi, Y.; Choi, H. J.; Kim, K. S. Observation of Tunable Band Gap and Anisotropic Dirac Semimetal State in Black Phosphorus. *Science* **2015,** *349* (6249), 723–726.

100. Buscema, M.; Groenendijk, D. J.; Blanter, S. I.; Steele, G. A.; van der Zant, H. S. J.; Castellanos-Gomez, A. Fast and Broadband Photoresponse of Few-Layer Black Phosphorus Field-Effect Transistors. *Nano. Lett.* **2014,** *14* (6), 3347–3352.

101. Hong, T.; Chamlagain, B.; Lin, W.; Chuang, H.-J.; Pan, M.; Zhou, Z.; Xu, Y.-Q. Polarized photocurrent response in black phosphorus field-effect transistors. *Nanoscale* **2014,** *6* (15), 8978–8983.

102. Wood, J. D.; Wells, S. A.; Jariwala, D.; Chen, K.-S.; Cho, E.; Sangwan, V. K.; Liu, X.; Lauhon, L. J.; Marks, T. J.; Hersam, M. C. Effective Passivation of Exfoliated Black Phosphorus Transistors against Ambient Degradation. *Nano. Lett.* **2014,** *14* (12), 6964–6970.

103. Cui, S.; Pu, H.; Wells, S. A.; Wen, Z.; Mao, S.; Chang, J.; Hersam, M. C.; Chen, J. Ultrahigh Sensitivity and Layer-Dependent Sensing Performance of Phosphorene-Based Gas Sensors. *Nat. Commun.* **2015,** *6*, 8632.

104. Park, C. M.; Sohn, H. J. Black Phosphorus and Its Composite for Lithium Rechargeable Batteries. *Advanced Materials* **2007,** *19* (18), 2465–2468.

105. Sun, J.; Lee, H.-W.; Pasta, M.; Yuan, H.; Zheng, G.; Sun, Y.; Li, Y.; Cui, Y. A phosphorene–Graphene Hybrid Material as a High-Capacity Anode for Sodium-Ion Batteries. *Nat. Nano* **2015,** *10* (11), 980–985.

Graphene composites with inorganic 2-D materials

5

CHAPTER

CHAPTER OUTLINE

5.1 Summary 103
5.2 Introduction 104
5.3 Graphene Composites With *h*-BN..105
    5.3.1 Design and Preparations 105
    5.3.2 Characterization 108
    5.3.3 Properties and Applications...110
5.4 Graphene Composites With MoS₂ 110
    5.4.1 Design and Preparation 110
    5.4.2 Characterization 112
    5.4.3 Properties and Applications...113
5.5 Composites With Other Inorganic Layered Materials...........................114
    5.5.1 Composites With TMDs 114
    5.5.2 Nongraphene Heterostructures..117
5.6 Conclusions...117
References 118

5.1 SUMMARY

Graphene hybrids with their inorganic counterparts make functional materials suitable for electronic applications. The design, synthesis, structural features, and applications of these heterostructures are discussed in this chapter. The inorganic layers can be tailored onto the graphene sheets by controlled growth to form so-called in-plane heterostructures. Alternatively, the layered material can be grown or transferred on top of each other, where they are stabilized by attractive dispersive interactions. These vertically stacked structures are fabricated by chemical vapor deposition (CVD) or micromechanical cleavage of the bulk crystals. Randomly oriented composites of graphene oxide (GO) or reduced graphene and inorganic layers are also prepared by solution-phase chemistry. Characterization of these hybrids is particularly challenging to prove the coexistence of graphene and the inorganic material in the same matrix. Elemental mapping and Raman microscopy techniques were explored here to address this challenge. The bandgap opening of graphene by hexagonal boron nitride (*h*-BN) and transition metal dichalcogenides (TMDs) are active materials for photodetectors, where graphene composites act as the photoelectrode. The inorganic sheets provide bandgap and photon absorption, whereas graphene imparts thermal/conductivity/chemical stability and mechanical strength. Details of the preparation, characterization, and properties of the hybrids have been tabulated to allow the reader to get a quick update in the field. We start with the discussion of h-BN/graphene hybrids followed by MoS_2/graphene. The reader may be able to figure out the similarity in synthesis and characterization of these materials, which can be later applied to heterostructures of other graphene/TMDs. Finally, a quick description of nongraphene heterostructures is given, which shows the postgraphene era of research on two-dimensional electronic materials.

5.2 INTRODUCTION

Hybrid materials of graphene and its inorganic family members display unique properties of both materials at the nano or atomic scale or they may also demonstrate synergy in addition to complementarity.[1] The layered structures of both graphene and its inorganic counterparts are beneficial for unifying these materials by maximum effective dispersive interactions.[2] These material combinations, when chosen appropriately, may introduce new attributes which are lacking in pure component phases and the whole could be larger than the sum of the parts. The interfaces of the stacked composite nanosheets are stabilized by interlayer van der Waals interactions and are called van der Waals heterostructures (Fig. 5.1A).[3,4] By controlled CVD, inorganic hexagonal patterns (such as h-BN or MoS_2) can be tailored onto graphene sheets by chemical bonding, called in-plane heterostructures (Fig. 5.1B).[3,5,6] Hybridizing graphene with other 2-D materials improves the electronic properties of graphene, and more importantly it could be used to tune

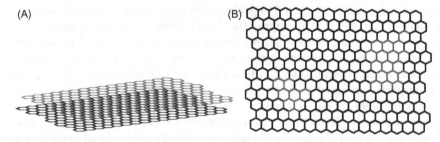

FIGURE 5.1

(A) Vertical heterostructure of graphene sheet (*black*) stacked with its inorganic analogue (*grey*). The sheets are held together by van der Waals forces, called van der Waals heterostructures. (B) In-plane heterostructure of graphene with inorganic 2-D materials embedded in the same sheet, where inorganic materials (in *grey*) are inserted into the openings of the hexagonal carbon lattice of graphene layer (in *black*).

the bandgap of graphene.[2,7] Zero bandgap of pristine graphene is not ideal for semiconductor devices, but this aspect is addressed by its composites, which is a major breakthrough for the construction of graphene-based electronics.[8−10]

The graphene bandgap is tuned by introducing strong interlayer or in-plane interactions between graphene and the inorganic layer with a known bandgap.[3,4] The first step towards preparation of tuned bandgap graphene composite is the design of a suitable synthetic method. The hybrid materials of graphene and inorganic nanosheets are generally prepared by two primary routes. For example, in one of the methods, both materials are prepared sequentially or simultaneously by directing the CVD or molecular beam epitaxy (MBE) processes.[11] In the second method, premade or exfoliated layers of the component phases are brought together by a transfer processes or by covalent chemical crosslinking.[12] CVD methods assure atomic precision of the heterostructures, while the physical/chemical contacts are simpler methods with the possibility of scaling up to large-scale production.[13]

Here, we discuss the nanostructures of graphene and inorganic layered materials with emphasis on the synthesis and characterization of the composites. The electronic properties of graphene composite materials were tested for performance in prototype devices. Boron nitride-based hybrids are discussed in detail first; these have been widely investigated for the last 10 years,[11] followed by MoS_2-based examples. Graphene hybrids with other TMDs and a few nongraphene heterostructures are reviewed towards the end of the chapter.

5.3 GRAPHENE COMPOSITES WITH *h*-BN

Controlled growth of graphene/*h*-BN nanostructures from its molecular precursors by the CVD method has been carried out.[11] It is done either by growing graphene on *h*-BN films or vice versa, depending on the original design. Two types of nanosheets result from this process: (1) in-plane structures where both graphene and *h*-BN domains are in the same sheet (Fig. 5.2A); and (2) vertically aligned structures where graphene and *h*-BN are stacked on top of each other (Fig. 5.2B).[11] Controlled introduction of precursors into the CVD chamber in a sequential manner was used to realize the above-mentioned hybrids.[14] Another route to achieve these composites is by conjugating exfoliated boron nitride and graphene (or functionalized graphene).[11] These methods and their features are discussed further.

5.3.1 DESIGN AND PREPARATIONS

Because of the structural similarity and isoelectronic character of *h*-BN with graphene, in-plane graphene/*h*-BN layers display strong lattice stability and these composites are also referred to as borocarbonitrides.[15] Some synthetic methods

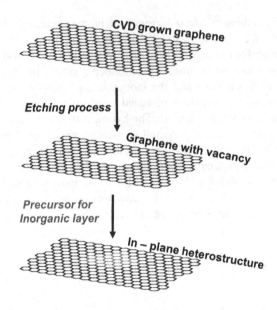

FIGURE 5.2

Sequence of processes to synthesize in-plane heterostructures. Shape selective etching of monolayer graphene can be achieved by photolithography or reactive ion etching[15] (Supporting substrates are avoided for illustration purposes).

for creating *h*-BN/graphene hybrids are listed in Table 5.1. The preparation steps involve the growth of graphene from a carbon source and that of *h*-BN from boron and nitrogen sources. For example, methane (CH_4) and ammonia borane (NH_3-BH_3) were used to grow in-plane graphene/*h*-BN structure on Cu substrates. The resulting structure had randomly distributed graphene and *h*-BN patterns and the percentage of carbon was tunable from 10% to 100%.[15] Postsynthetic modification of CVD-grown graphene layers by boron nitride deposition was used to realize shape control for in-plane heterostructures of these materials.[18] Using photolithography or reactive ion etching, parts of graphene domains were etched out from the supporting surface and *h*-BN was grown subsequently at these areas to yield composite nanostructures (Fig. 5.2).[18] Complete conversion of graphene to boron nitride layers was also reported by a CVD method, where near-perfect graphene films were exposed to ammonia borane at 1000°C.[25] Interestingly, all carbon atoms in graphene were completely replaced by B and N after 2 hours of exposure. Similar control over CVD conditions developed to form vertical stacks of graphene and *h*-BN composites.

The low lattice mismatch (1.7%) between graphene and *h*-BN, when compared to commonly used metal supports in CVD (Cu or Ni), was exploited to grow nearly defect-free, large-area graphene domains on *h*-BN.[2] The sequential CVD growth method was also adopted to fabricate stacked layers of graphene

Table 5.1 Synthesis, Characterization, and Properties of Some Graphene/
h-BN Heterostructures

Method	Precursors	Major Characterization Techniques	Remarks
Chemical vapor deposition (CVD)[16]	*h*-BN flakes and CVD-grown graphene	Raman spectroscopy, optical microscopy, and atomic force microscopy (AFM),	Charge carrier mobility of $\sim$ 2250 cm^2/V/s
CVD[17]	Borazine and methane	Scanning electron microscopy (SEM), X-ray photoelectron spectroscopy (XPS), and high-resolution transmission electron microscopy (HR-TEM)	Fabricated in a field effect transistor
CVD and photolithography[18]	Ammonia borane and methane	Raman spectroscopy, optical microscopy, SEM, AFM, and TEM	Fabricated in a field effect transistor
CVD[15]	Ammonia borane and methane	Raman spectroscopy, AFM, HR-TEM, electron energy-loss spectroscopy (EELS), and XPS	Conductivity increases with carbon content ($10 - 3$ Ω.cm at 94% carbon content) Bandgap tuning with C content
CVD[19]	Ammonia borane and benzoic acid	Scanning tunneling microscopy (STM), SEM, AFM, Raman spectroscopy, XPS, and Auger electron spectroscopy	Temperature control to make in-plane or vertically stacked composites
Low-pressure CVD[20]	Ammonia borane and methane	AFM (Kelvin probe), Raman spectroscopy, SEM, and scanning tunneling electron microscopy–EELS	"AA-like" stacks of graphene/*h*-BN
Molecular beam epitaxy[21]	B_2O_3 powder and acetylene gas	Raman spectroscopy, SEM, XPS, and TEM	—
Molecular beam epitaxy[22]	Graphite, elemental B and N	Raman Spectroscopy, AFM, XPS, and grazing incidence X-ray diffraction	Ni/MgO(111) substrate
Chemical crosslinking[23]	Synthesized boron nitride and carboxylated graphene	Raman spectroscopy, FTIR, XPS, solid-state NMR (^{13}C), TEM, and powder XRD	Supercapacitor, catalyst for oxygen reduction
Liquid-phase exfoliation[24]	Graphite and *h*-BN	SEM, TEM-EDX, AFM, FTIR, and UV-vis spectroscopy	Bulk crystals were previously expanded in conc. HNO_3 and H_2SO_4

and boron nitride, where the carbon source from graphene on the substrate followed by the introduction of *h*-BN precursor. For example, *n*-hexane and ammonia borane were used as precursors for graphene and *h*-BN, respectively, to grow these stacks.[26] Similarly, highly oriented pyrolytic graphite or mechanically exfoliated graphene can be used to grow *h*-BN by CVD, which also results in layer-by-layer assembly of graphene and *h*-BN, with total thickness varying from 8 to 20 nm.[26] As an alternative to CVD, the MBE technique was developed to generate graphene/*h*-BN nanolayer stacks on Ni/MgO (111) surface.[22] In this process, carbon was deposited by annealing the substrate at 850°C under carbon vapor generated from an electron beam evaporated graphite surface. *h*-BN was coated on this surface by exposure to nitrogen plasma and a boron beam from a high-temperature effusion cell and a plasma source.[22] All these processes with atomic control over film growth lack scalability but transferring the composites onto other substrates can be challenging.[5] In contrast, the top-down methods, described below, offer scalable synthetic approaches to make bulk quantities of *h*-BN/graphene heterostuctures.

Chemical crosslinking of carboxylated graphene and boron nitride was demonstrated as a facile route to prepare nanocomposites with varying weight ratios of BN to graphene (0.25, 0.5, 0.75).[23] The amide coupling between carboxylic groups on graphene and the amine edge groups in boron nitride (previously synthesized from boric acid and urea) by ethyl-3-(3-dimethylaminopropyl) carbodiimide was used to make the layered assembly.[23] This approach offers gram-scale synthesis of the hybrid material, although the individual sheets are randomly oriented. Simultaneous liquid-phase exfoliation of graphite and *h*-BN in *N*,*N*-dimethylformamide by ultrasonication, after expansion of its bulk crystals by treating with conc. HNO_3/H_2SO_4 is a promising route for large-scale preparation of the corresponding composite materials with particular compositions.[24] This method yielded hybrid sheets of about 100 nm thickness. However, control over the layering, sheet thickness, and relative compositions of the two phases requires further work. Some synthesis methods of the graphene and *h*-BN heterostructures are summarized in Table 5.1. Next, we discuss key methods used for the characterization of this hybrid material.

5.3.2 CHARACTERIZATION

The major aim of characterizing graphene/*h*-BN composites is to evaluate the coexistence of both the graphene and *h*-BN domains in the same matrix. Spectroscopic techniques including absorption, infrared, X-ray photoelectron and Raman spectroscopy, and various microscopic techniques were employed for the primary characterization of the individual domains in the composite matrix. Detailed characterization methods of pure phases of graphene and boron nitride are discussed in Chapter 3, Characterization techniques for graphene and Chapter 4, Inorganic analogues of graphene: synthesis, characterization, and applications, and these techniques are also suitable for the characterization of the heterostructures.

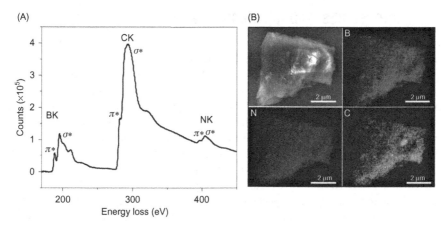

FIGURE 5.3

(A) Electron energy-loss spectra of graphene/*h*-BN in-plane hybrid nanosheet shows the energy-loss corresponding to C, B, and N atoms in C−C, B−N, and B−C bonding (©Nature Publishing Group).[15] (B) Elemental mapping of graphene/*h*-BN composites shows the presence of boron (*top right*), nitrogen (*bottom left*), and carbon (*bottom right*) in the same flake (© John Wiley and sons).[23]

Nanoscale elemental mapping techniques are particularly useful to examine these heterogeneous matrices to elucidate their chemical composition.[27]

High-resolution TEM or scanning tunneling electron microscopy (STEM) with energy-dispersive electron (or X-ray) spectroscopy (EDX) capability or electron energy-loss spectroscopy (EELS) are widely used to characterize both in-plane and vertically stacked graphene/*h*-BN nanostructures.[15,20] Loss of electron energy due to absorption of elements on exposure to electron beam can be used to examine the composition of materials under the electron beam.[28] EELS map and corresponding spectra of in-plane graphene/*h*-BN structures showed signatures of both carbon and boron atoms, on nanometer scale, supporting their coexistence in the samples.[15] Fig. 5.3A shows the spectra of in-plane graphene−*h*-BN layers synthesized by the CVD method, where the peaks corresponding to B ($\sim$191 eV), C ($\sim$285 eV), and N ($\sim$398 eV) are present, and the data prove the coexistence of boron, carbon, and nitrogen in the nanosheets.[15] Similarly, the EDX map of composites of carboxylated graphene and boron nitride films by chemical crosslinking shows simultaneous appearance of boron, nitrogen, and carbon (Fig. 5.3B) on the same particle.[23] Atomic force microscopy with in-built Raman spectra acquisition is another useful tool to map the characteristic resonance bands of graphene and *h*-BN in the same flake.[22] However, more widely available tools such as optical microscopy and SEMs can also be used to qualitatively assess the success of growth of graphene or *h*-BN films, by looking at the contrast differences.[15,20] Practical applications of such thin layers of the composites or heterostructures are discussed next.

5.3.3 PROPERTIES AND APPLICATIONS

Electronic properties and applications of graphene/h-BN hybrids are promising for dramatic improvements in device design and efficiency. The bandgap tuning of graphene (zero bandgap) with h-BN (~ 5.8 eV), e.g., broadens the scope for semi-conducting applications.[15] Up to 4 eV opening of graphene bandgap was accomplished by in-plane modification of h-BN on graphene (at 64% C).[15] Conversely, carbon inclusion to h-BN domains increases the electron conductivity of h-BN, which is an insulator in its pristine form.[15] The induced bandgap in graphene was effectively used to fabricate field effect transistors of graphene/h-BN films with carrier mobilities ranging from 190 to 2000 cm^2/V/second (pure graphene carrier mobility is 2×10^5 cm^2/V/second).[29,30] Composites of chemically crosslinked graphene and boron nitride materials improved energy storage efficiency.[23] The graphene/h-BN supercapacitors show maximum specific capacitance of ~ 240 F/g.[23] The capacitance increased steadily with an increase of h-BN mole fraction in the composite, which has been attributed to the higher surface area and microporosity of the composite over the pure phases.[23] Likewise, graphene/h-BN composites also improved the performance of Li-ion batteries,[31] where the discharge capacity was increased with higher C content from 1000 to 1900 mAh/g.[32] These properties and applications of hybrids are listed in Table 5.1. Another class of widely studied heterostructures of graphene is with MoS$_2$, which is reviewed next.

5.4 GRAPHENE COMPOSITES WITH MoS$_2$

Molybdenum disulfide (MoS$_2$) is a member of the TMDs, which have a thickness-dependent bandgap, which is in a suitable range for applications in semiconductor devices. High tensile strength and chemical/thermal stability of graphene can be combined with unique optoelectronic and photovoltaic nature of MoS$_2$ for improved performance of the composite over the parent components. Similar to the graphene/h-BN heterostructures, graphene/MoS$_2$ hybrids are also prepared by controlled CVD methods. The synthetic design and strategies of these composites are discussed below.

5.4.1 DESIGN AND PREPARATION

The CVD technique offers highly defined structures and this is often achieved by controlling the introduction of the precursor at predefined levels and intervals. Commonly used Mo sources for MoS$_2$ synthesis include MoO$_3$, ammonium tetrathiomolybdate ((NH$_4$)$_2$MoS$_4$), and sulfur. Alternatively, MoS$_2$ thin layers are transferred onto the graphene films to form the composites of specific compositions of the two phases.[8] Chemical synthesis of MoS$_2$ with GO or reduced GO is also used to form high surface area composites which are primarily targeted for Li-ion battery applications.[2] A review of some of the synthetic approaches is listed in Table 5.2,

Table 5.2 Synthesis, Characterization, and Properties of Some GrapheneMoS$_2$ Heterostructures

Method	Precursors	Major Characterization Techniques	Remarks
Thermal annealing and CVD[33]	Silicon carbide, ammonium heptamolybdate and sulfur vapor	Raman Spectroscopy, scanning Transmission electron Microscopy (STEM), photoluminescence spectroscopy, XPS and angle resolved photoemission spectroscopy	• Interlayer separation of 3.4 ± 0.1 Å nm • Optical bandgap of 1.8 eV
Ambient-pressure CVD[34]	MoO$_3$, sulfur, and CVD-grown graphene	Optical microscopy, Raman spectroscopy, photoluminescence spectroscopy, HR-TEM-EDX, and AFM	• Perylene-3,4,9,10-tetracarboxylic acid tetrapotassium salt (PTAS) was used as a seed • MoS$_2$ layers were physically transferred to graphene substrate
Ambient-pressure CVD[35]	CH$_4$, MoO$_3$, and sulfur	SEM, AFM, Raman spectroscopy and photoluminescence spectroscopy	• Vertically aligned morphology • Illustrated as visible light photodetector
CVD and molecular beam epitaxy (MBE)[36]	CH$_4$, MoO$_3$, and sulfur	SEM, HR-TEM, Raman spectroscopy, and photoluminescence spectroscopy	• Graphene from CVD method and MoS$_2$ by MBE • Electrochemical bubbling technique to transfer thin layers
Low-pressure CVD[37]	CH$_4$, MoCl$_5$, and sulfur	HR-TEM, AFM, Raman spectroscopy, and photoluminescence spectroscopy	• Postgrowth annealing under sulfur vapor increased the crystallinity and optoelectronic properties
CVD[38]	Ethanol and (NH$_4$)$_2$MoS$_4$	XRD, SEM, TEM, and confocal Raman microscopy	• Ethanol as a carbon source for 3-D graphene network[39] • Reversible energy density of 877 mAh/g in Li-ion battery
Hydrothermal synthesis[40]	Bulk MoS$_2$ and graphene oxide (GO)	XRD, SEM, HR-TEM, BET surface area measurement, and	• MoS$_2$ exfoliation by electrochemical intercalation with Li

(Continued)

Table 5.2 Synthesis, Characterization, and Properties of Some
GrapheneMoS$_2$ Heterostructures *Continued*

Method	Precursors	Major Characterization Techniques	Remarks
		thermogravimetric analysis (TGA)	• GO was reduced to rGO • Reversible charging with energy density of 1200 mAh/g in Li-ion battery
Lithiation-exfoliation[41]	Bulk MoS$_2$ and reduced graphene oxide (rGO)	XRD, XPS, SEM, HR-TEM, Raman spectroscopy, and elemental analysis	• Few-layer MoS$_2$ (1–3 layers) • Energy density of 1200 mAh/g at current density of 100 mA/g
Freeze drying/ thermal annealing[42]	Graphene aerogel[43] and (NH$_4$)$_2$MoS$_4$	XPS, SEM-EDX, HR-TEM, and Raman spectroscopy	• Graphene aerogel from GO, formaldehyde and resorcinol • NO$_2$ detection as low as 50 ppb

which shows the similarity with the methods described for graphene/*h*-BN hetero-structures. A promising method not listed here is the covalent linking of the two phases with specific linkage chemistry and functionalities.

5.4.2 CHARACTERIZATION

Most of the spectroscopic and microscopic techniques used for the characterization of 2-D materials reported in Chapter 3, Characterization techniques for graphene and Chapter 4, Inorganic analogues of graphene: synthesis, characterization, and applications, are also applicable for these composites. Distinctively, the photoluminescence property[44] of MoS$_2$ nanolayers gives another handle to identify its coexistence with graphene. Elemental mapping techniques, such as EELS and EDX, are required to prove the chemical composition at nanoscale. The photoluminescence intensity of MoS$_2$ can also be used to map the grown layers, in conjunction with the Raman map and optical image as shown in Fig. 5.4. Strong optical absorption of MoS$_2$ at ~600 and ~675 nm, where graphene (or GO) is optically transparent, is exploited to prove the presence of these layers in solution.[40] The characterization techniques adopted for these composites are listed in Table 5.2. Next, the unique properties and demonstrated applications of these hybrid materials are discussed.

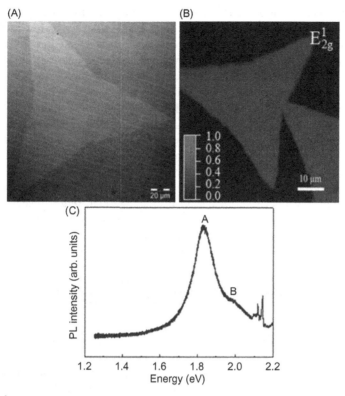

FIGURE 5.4

Visualization of CVD-grown MoS_2 on graphene using (A) optical imaging, (B) Raman microscopy (385 cm^{-1}), and (C) photoluminescence spectra (© American Chemical Society).[33]

5.4.3 PROPERTIES AND APPLICATIONS

Graphene/MoS_2 hybrids find specific applications in electronic devices mostly because of the introduction of bandgap in graphene which corresponds to the visible light wavelengths. MoS_2 with a visible range bandgap of 1.2−1.9 eV was fabricated with graphene to make photo-responsive or photo-detecting devices.[9,45] The photodetection works based on the current generation by the MoS_2 surface upon illumination (photocurrent). Graphene is not suitable for photocurrent generation because of a lack of bandgap.[9] The photocurrent generation and annihilation is in a millisecond timeframe (ON/OFF) even under microwatt power illumination.[40] For instance, a photo-detecting device fabricated using graphene/MoS_2 thin layers showed 5×10^8 A/W responsivity ($\lambda = 635$ nm, LED power = 6.4 fW/μm^2) at room temperature, which is highest for a graphene-based photodetector while the response of pristine graphene is $\leq 1 \times 10^{-2}$ A/W.[46] The photo-responsive nature of these hybrids was also applied for sensing applications, where nitrogen dioxide was selectively detected at low levels, as low as 50 ppb.[40] The electron-accepting nature of NO_2 causes electron extraction from the graphene/MoS_2 composite, upon adsorption, and causes a decrease in resistance, which generates a signal output.[40] Surface plasmon resonance (SPR)-based biosensors, which are fabricated using graphene/MoS_2 layers (Fig. 5.5A), show a 500-fold increase in phase sensitivity when compared to the traditional gold film (Fig. 5.5B).[47] This has been attributed to the high optical absorption of MoS_2 phase and efficient energy transfer to the graphite layer.[46] Composites of GO and MoS_2 were used as an anodic material in Li-ion batteries, because of its Li^+ intercalation capacity and high surface area.[38–40] These heterostructures achieved energy densities up to 1200 mAh/g, in comparison to commercial lithium titanate (LTO) batteries, which have a capacity of only $\sim$200 mAh/g.[48] Many important applications for this class of materials are listed in Table 5.2. Next, we discuss the hybrid materials of graphene with other TMDs.

5.5 COMPOSITES WITH OTHER INORGANIC LAYERED MATERIALS

Although h-BN and MoS_2 are popular for making graphene heterostructures, other TMDs also form interesting hybrids with graphene to target specific applications. Another motivation behind the construction of these other structures is to further tune graphene properties. These hybrids are also finding applications in energy storage, catalysis, and sensing.[2] A short list of such composites and their features is presented in Table 5.3.

5.5.1 COMPOSITES WITH TMDS

The direct bandgap and photosensitivity of TMDs are mostly explored in the hybrid structures of graphene. The differences in bandgap between TMDs make it fit for photovoltaic devices working under various optical wavelengths.[6] The availability of bulk crystals of these materials is also an advantage for easy fabrication of the devices. After MoS_2/graphene, heterostructures of WS_2/graphene are the most studied. These are synthesized by CVD from molecular precursors or by simple mechanical cleavage from their bulk crystals.[61] Highly sensitive photodetectors are fabricated from these junctions by harnessing the electronic coupling between graphene and WS_2. The bandgap of a few-layer material is at $\sim$2.1 eV, which is suitable for visible light detection, and the devices show current on/off ratios of $\sim$10^6.[51,52] Similarly other TMDs such as WSe_2 and $MoSe_2$ are also studied to form functional hybrids with graphene and the prototype devices show better performance as photodetectors than pristine graphene-based systems (Table 5.3). Synthetic methods and properties of some TMD/graphene hybrids are listed in Table 5.3.

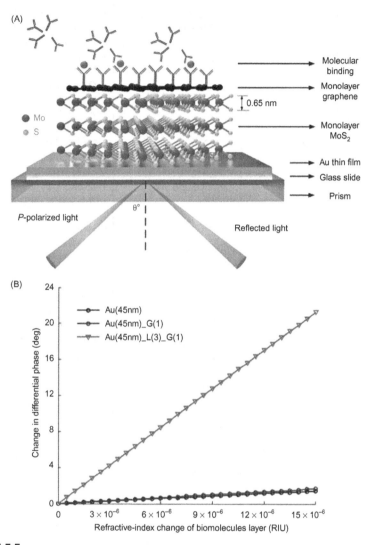

FIGURE 5.5

(A) A high-profile biological application of MoS_2/graphene heterostructure is demonstrated by integrating it to surface plasmon resonance (SPR) configuration for detection of biomolecules. (B) Five hundred times increase in output signal, which changes in the differential phase of the reflected light, after the modification with MoS_2/graphene heterostructure is observed, in comparison to bare gold or graphene on a gold surface (© Elsevier).[47]

Table 5.3 Example of Graphene/TMD Heterostructures and Their Properties

Hybrid	Method	Precursors	Remarks
WS_2/ rGO[49,50]	Hydrothermal	WCl_6, thioacetamide (C_2H_5NS) and GO	• Supercapacitor of the hybrid shows energy density of 49 Wh/kg • Fabricated in a field effect transistor (FET) • Orbital overlapping between WS_2 and rGO was proposed
WS_2/ graphene[51]	CVD + thermal annealing	Acetylene and ammonium tetrathiotungstate (($NH_4)_2WS_4$)	• Catalyst for hydrogen evolution reaction • No degradation after 2000 catalytic cycles
WS_2/ graphene[52]	Micromechanical cleavage	Graphite and WS_2 crystal	• 4- or 5-layer WS_2 was sandwiched between graphene and h-BN • Highest on/off ratio of 1 × 10^6 at room temperature (FET)
WS_2/ graphene[53]	Micromechanical cleavage + CVD	Methane and WS_2 crystal	• Photoluminescence of WS_2 nanosheet was quenched by graphene • On/off ratio of 10^6 n-type and 10^2 p-type (photodetecting device)
WSe_2/ graphene[54]	CVD	Methane, tungsten hexacarbonyl ($W(CO)_6$) and dimethylselenide (DMSe)	• Photoluminescence of TMD was quenched by graphene
WSe_2/ graphene[55]	Micromechanical cleavage	Graphite and WSe_2 crystal	• Band alignment of WSe_2 and graphene • > 10 layers of WSe_2
WSe_2/ graphene[56]	CVD and micromechanical cleavage	CVD-grown graphene and WSe_2 crystals	• On/off-current ratio of 3 × 10^4 at room temperature (Barristor device)
$MoSe_2$/ graphene[57]	CVD	Methane, MoO_3, and selenium powder	• Quenching of photoluminescence due to charge transfer from $MoSe_2$ to graphene
$MoSe_2$/ graphene[58]	Micromechanical cleavage	Graphite and $MoSe_2$ crystal	• 14−81 layers of $MoSe_2$ • On/off-current ratio of 10^5 at 300 K (FET)
$MoTe_2$/ graphene[59]	Micromechanical cleavage	CVD-grown bilayer graphene and $MoTe_2$ crystal	• ∼10 layer $MoTe_2$ • $MoTe_2$ band gap is ∼1.1 eV (infrared region) • Current on/off ratio ∼10^5 at 77 K (FET)
Bi_2Se_3/ graphene[60]	Micromechanical cleavage/CVD	Graphite and Bi_2Se_3	• Catalyst-free CVD for Bi_2Se_3 to form 1 to 9 layers

5.5.2 NONGRAPHENE HETEROSTRUCTURES

Other van der Waals heterostructures that do not involve graphene are also known and indicate interesting applications, which deserve mentioning in the context of graphene, as these also belong to the same family of 2-D materials. These other heterostructures are fabricated using different combinations of layered material stacks such as MoS_2/h-BN,[62,63] MoS_2/black phosphorus,[64] MoS_2/WSe_2,[65] MoS_2/WS_2,[66] MoS_2/SnS_2,[67] WS_2/SnS_2,[65] $WS_2/MoSe_2$,[63] WSe_2/SnS_2,[65] and $WSe_2/MoSe_2$.[68] These hybrids are interesting for photovoltaic applications because their bandgap can be tuned systematically and their photoluminescence properties can be controlled at will. Detailed discussions of these hybrids are beyond the scope of this book, but interested readers may want to read some of the many references cited above.

5.6 CONCLUSIONS

Graphene as a primary two-dimensional family member is a promising candidate for electronic applications because of its high carrier mobility, electrical conductivity, mechanical strength, thermochemical stability, etc.[69] But its semimetallic nature (zero bandgap) is not a fit for semiconducting applications. On the other hand, inorganic graphene-like materials possess direct banddap, especially transition metal dichalogenides, and they are promising for photovoltaic devices, especially when combined with the advantages of grpahene.[6] Hybridizing graphene with its inorganic counterparts provides opportunities to systematically engineer the bandgap of graphene to magnitudes that are more suitable for the semiconductor industry.[3] Electronic and orbital coupling by in-plane or vertical stacks of graphene with these 2-D materials resulted in raising the graphene bandgap up to 2 eV.[7] Remarkable photosensitivity of these hybrids is reported ($\sim 10^6$ on/off ratio), which cannot be achieved by individual materials, proving the synergy by combining graphene and the inorganic layers.[2] CVD is advantageous in fabricating shape-controlled heterostructures, but chemical crosslinking or micromechanical cleavage from bulk crystals is also interesting. Controlled fabrication of the heterostructures using CVD takes several steps and transfer of the material to flexible substrates on a large scale is still challenging. The ambient stability of the CVD-synthesized TMDs is also questionable because of their vulnerability to air oxidation.[70] Micromechanically cleaved crystals are also used for fabrication, but reproducibility in the thickness of the material is a challenge, which affects the device performance.

Sophisticated characterization techniques are often required in this field because of the chemical complexity of the composites, and this is because the structure and morphology of different domains are often indistinguishable by standard imaging techniques. Elemental mapping, Raman mapping, or other methods of contrasting different domains of the heterostructures is essential to ascertain the presence of both types of nanosheet in the composite. The use of liquid-phase exfoliation for simple and scalable synthesis of these heterostructures still needs to be explored for the preparation of composite structures, which may have a broader impact for device fabrication. Controlled synthesis of these hybrids on a large scale, while ensuring reproducibility, is an important currently unmet challenge.

REFERENCES

1. Hong, H.; Liu, C.; Cao, T.; Jin, C.; Wang, S.; Wang, F.; Liu, K. Interfacial Engineering of Van der Waals Coupled 2D Layered Materials. *Adv. Mater. Interfaces* **2017,** *4* (9), 1601054 (article number).
2. Niu, T.; Li, A. From Two-Dimensional Materials to Heterostructures. *Prog. Surface Sci.* **2015,** *90* (1), 21–45.
3. Geim, A. K.; Grigorieva, I. V. Van der Waals Heterostructures. *Nature* **2013,** *499* (7459), 419–425.
4. Novoselov, K.; Mishchenko, A.; Carvalho, A.; Neto, A. C. 2D Materials and Van Der Waals Heterostructures. *Science* **2016,** *353* (6298), aac9439.
5. Chen, X.; Wu, B.; Liu, Y. Direct Preparation of High Quality Graphene on Dielectric Substrates. *Chem. Soc. Rev.* **2016,** *45* (8), 2057–2074.
6. Wang, H.; Liu, F.; Fu, W.; Fang, Z.; Zhou, W.; Liu, Z. Two-Dimensional Heterostructures: Fabrication, Characterization, and Application. *Nanoscale* **2014,** *6* (21), 12250–12272.
7. Zeng, Q.; Wang, H.; Fu, W.; Gong, Y.; Zhou, W.; Ajayan, P. M.; Lou, J.; Liu, Z. Band Engineering for Novel Two-Dimensional Atomic Layers. *Small* **2015,** *11* (16), 1868–1884.
8. Tan, C.; Zhang, H. Two-Dimensional Transition Metal Dichalcogenide Nanosheet-Based Composites. *Chem. Soc. Rev.* **2015,** *44* (9), 2713–2731.
9. Koppens, F.; Mueller, T.; Avouris, P.; Ferrari, A.; Vitiello, M.; Polini, M. Photodetectors Based on Graphene, Other Two-Dimensional Materials and Hybrid Systems. *Nat. Nanotechnol.* **2014,** *9* (10), 780–793.
10. Xie, C.; Mak, C.; Tao, X.; Yan, F. Photodetectors Based on Two-Dimensional Layered Materials Beyond Graphene. *Adv. Funct. Mater.* **2016.** Available from: http://dx.doi.org/10.1002/ adfm.201603886.
11. Li, Q.; Liu, M.; Zhang, Y.; Liu, Z. Hexagonal Boron Nitride–Graphene Heterostructures: Synthesis and Interfacial Properties. *Small* **2016,** *12* (1), 32–50.
12. Akinwande, D.; Petrone, N.; Hone, J. Two-Dimensional Flexible Nanoelectronics. *Nat. Commun.* **2014,** *5*, 5678.
13. Yin, J.; Li, J.; Hang, Y.; Yu, J.; Tai, G.; Li, X.; Zhang, Z.; Guo, W. Boron Nitride Nanostructures: Fabrication, Functionalization and Applications. *Small* **2016,** *12* (22), 2942–2968.
14. Wu, Q.; Wongwiriyapan, W.; Park, J.-H.; Park, S.; Jung, S. J.; Jeong, T.; Lee, S.; Lee, Y. H.; Song, Y. J. In Situ Chemical Vapor Deposition of Graphene and Hexagonal Boron Nitride Heterostructures. *Curr. Appl. Phys.* **2016,** *16* (9), 1175–1191.

15. Ci, L.; Song, L.; Jin, C.; Jariwala, D.; Wu, D.; Li, Y.; Srivastava, A.; Wang, Z.; Storr, K.; Balicas, L. Atomic Layers of Hybridized Boron Nitride and Graphene Domains. *Nat. Mater.* **2010,** *9* (5), 430–435.
16. Wang, M.; Jang, S. K.; Song, Y. J.; Lee, S. CVD Growth of Graphene Under Exfoliated Hexagonal Boron Nitride for Vertical Hybrid Structures. *Mater. Res. Bull.* **2015,** *61*, 226–230.
17. Wu, Q.; Jang, S. K.; Park, S.; Jung, S. J.; Suh, H.; Lee, Y. H.; Lee, S.; Song, Y. J. In Situ Synthesis of a Large Area Boron Nitride/Graphene Monolayer/Boron Nitride Film by Chemical Vapor Deposition. *Nanoscale* **2015,** *7* (17), 7574–7579.
18. Levendorf, M. P.; Kim, C.-J.; Brown, L.; Huang, P. Y.; Havener, R. W.; Muller, D. A.; Park, J. Graphene and Boron Nitride Lateral Heterostructures for Atomically Thin Circuitry. *Nature* **2012,** *488* (7413), 627–632.
19. Gao, T.; Song, X.; Du, H.; Nie, Y.; Chen, Y.; Ji, Q.; Sun, J.; Yang, Y.; Zhang, Y.; Liu, Z. Temperature-Triggered Chemical Switching Growth of In-Plane and Vertically Stacked Graphene-Boron Nitride Heterostructures. *Nat. Commun.* **2015,** *6*, 6835.
20. Kim, S. M.; Hsu, A.; Araujo, P.; Lee, Y.-H.; Palacios, T. S.; Dresselhaus, M.; Idrobo, J.-C.; Kim, K. K.; Kong, J. Synthesis of Patched or Stacked Graphene and hBN Flakes: A Route to Hybrid Structure Discovery. *Nano Lett.* **2013,** *13* (3), 933–941.
21. Zuo, Z.; Xu, Z.; Zheng, R.; Khanaki, A.; Zheng, J.-G.; Liu, J. In-Situ Epitaxial Growth of Graphene/h-BN Van Der Waals Heterostructures by Molecular Beam Epitaxy. *Sci. Rep.* **2015,** *5*, 14760.
22. Wofford, J. M.; Nakhaie, S.; Krause, T.; Liu, X.; Ramsteiner, M.; Hanke, M.; Riechert, H.; Lopes, J. M. J. A Hybrid MBE-Based Growth Method for Large-Area Synthesis of Stacked Hexagonal Boron Nitride/Graphene Heterostructures. *Sci. Rep.* **2017,** *7*, 43644.
23. Kumar, R.; Gopalakrishnan, K.; Ahmad, I.; Rao, C. BN–Graphene Composites Generated by Covalent Cross-Linking with Organic Linkers. *Adv. Funct. Mater.* **2015,** *25* (37), 5910–5917.
24. Güler, Ö.; Güler, S. H. Production of Graphene–Boron Nitride Hybrid Nanosheets by Liquid-Phase Exfoliation. *Optik-Int. J. Light Electron Optics* **2016,** *127* (11), 4630–4634.
25. Gong, Y.; Shi, G.; Zhang, Z.; Zhou, W.; Jung, J.; Gao, W.; Ma, L.; Yang, Y.; Yang, S.; You, G. Direct Chemical Conversion of Graphene to Boron-and Nitrogen-and Carbon-Containing Atomic Layers. *Nat. Commun.* **2014,** *5*, 3193.
26. Liu, Z.; Song, L.; Zhao, S.; Huang, J.; Ma, L.; Zhang, J.; Lou, J.; Ajayan, P. M. Direct Growth of Graphene/Hexagonal Boron Nitride Stacked Layers. *Nano Lett.* **2011,** *11* (5), 2032–2037.
27. Haigh, S. J.; Rooney, A. P.; Prestat, E.; Withers, F.; Del Pozo Zamudio, O.; Mishchenko, A.; Gholinia, A.; Watanabe, K.; Taniguchi, T.; Tartakovskii, A. I., et al. Cross Sectional STEM Imaging and Analysis of Multilayered Two Dimensional Crystal Heterostructure Devices. *Microscopy Microanal.* **2015,** *21* (S3), 107–108.
28. Hillier, J.; Baker, R. F. Microanalysis by Means of Electrons. *J. Appl.Phys.* **1944,** *15* (9), 663–675.
29. Chang, C.-K.; Kataria, S.; Kuo, C.-C.; Ganguly, A.; Wang, B.-Y.; Hwang, J.-Y.; Huang, K.-J.; Yang, W.-H.; Wang, S.-B.; Chuang, C.-H., et al. Band Gap Engineering of Chemical Vapor Deposited Graphene by in Situ BN Doping. *ACS Nano* **2013,** *7* (2), 1333–1341.

30. Rao, C. N. R.; Gopalakrishnan, K. Borocarbonitrides, $B_xC_yN_z$: Synthesis, Characterization, and Properties with Potential Applications. *ACS Appl. Mater. Interfaces* **2016**.

31. Li, H.; Tay, R. Y.; Tsang, S. H.; Liu, W.; Teo, E. H. T. Reduced Graphene Oxide/ Boron Nitride Composite Film as a Novel Binder-Free Anode for Lithium Ion Batteries With Enhanced Performances. *Electrochim. Acta* **2015**, *166*, 197−205.

32. Sen, S.; Moses, K.; Bhattacharyya, A. J.; Rao, C. Excellent Performance of Few-Layer Borocarbonitrides as Anode Materials in Lithium-Ion Batteries. *Chem. Asian J.* **2014**, *9* (1), 100−103.

33. Pierucci, D.; Henck, H.; Avila, J.; Balan, A.; Naylor, C. H.; Patriarche, G.; Dappe, Y. J.; Silly, M. G.; Sirotti, F.; Johnson, A. C. Band Alignment and Minigaps in Monolayer MoS_2-Graphene Van Der Waals Heterostructures. *Nano Lett.* **2016**, *16* (7), 4054−4061.

34. Lee, Y.-H.; Yu, L.; Wang, H.; Fang, W.; Ling, X.; Shi, Y.; Lin, C.-T.; Huang, J.-K.; Chang, M.-T.; Chang, C.-S. Synthesis and Transfer of Single-Layer Transition Metal Disulfides on Diverse Surfaces. *Nano Lett.* **2013**, *13* (4), 1852−1857.

35. Yunus, R. M.; Endo, H.; Tsuji, M.; Ago, H. Vertical Heterostructures of MoS_2 and Graphene Nanoribbons Grown by Two-Step Chemical Vapor Deposition for High-Gain Photodetectors. *Phys.Chem. Chem. Phys.* **2015**, *17* (38), 25210−25215.

36. Wan, W.; Li, X.; Li, X.; Xu, B.; Zhan, L.; Zhao, Z.; Zhang, P.; Wu, S.; Zhu, Z.-Z.; Huang, H. Interlayer Coupling of a Direct Van Der Waals Epitaxial MoS_2/Graphene Heterostructure. *RSC Adv.* **2016**, *6* (1), 323−330.

37. Wu, C.-R.; Liao, K.-C.; Wu, C.-H.; Lin, S.-Y. Luminescence Enhancement and Enlarged Dirac Point Shift of MoS_2/Graphene Hetero-Structure Photodetectors with Postgrowth Annealing Treatment. *IEEE J. Selected Topics Quant. Electron.* **2017**, *23* (1), 1−5.

38. Cao, X.; Shi, Y.; Shi, W.; Rui, X.; Yan, Q.; Kong, J.; Zhang, H. Preparation of MoS_2-Coated Three-Dimensional Graphene Networks for High-Performance Anode Material in Lithium-Ion Batteries. *Small* **2013**, *9* (20), 3433−3438.

39. Cao, X.; Shi, Y.; Shi, W.; Lu, G.; Huang, X.; Yan, Q.; Zhang, Q.; Zhang, H. Preparation of Novel 3D Graphene Networks for Supercapacitor Applications. *Small* **2011**, *7* (22), 3163−3168.

40. Jiang, L.; Lin, B.; Li, X.; Song, X.; Xia, H.; Li, L.; Zeng, H. Monolayer MoS_2−Graphene Hybrid Aerogels with Controllable Porosity for Lithium-Ion Batteries with High Reversible Capacity. *ACS Appl. Mater. Interfaces* **2016**, *8* (4), 2680−2687.

41. Jing, Y.; Ortiz-Quiles, E. O.; Cabrera, C. R.; Chen, Z.; Zhou, Z. Layer-by-Layer Hybrids of MoS_2 and Reduced Graphene Oxide for Lithium Ion Batteries. *Electrochim. Acta* **2014**, *147*, 392−400.

42. Long, H.; Harley-Trochimczyk, A.; Pham, T.; Tang, Z.; Shi, T.; Zettl, A.; Carraro, C.; Worsley, M. A.; Maboudian, R. High Surface Area MoS_2/Graphene Hybrid Aerogel for Ultrasensitive NO_2 Detection. *Adv. Funct. Mater.* **2016**, *26* (28), 5158−5165.

43. Worsley, M. A.; Pauzauskie, P. J.; Olson, T. Y.; Biener, J.; Satcher, J. H., Jr; Baumann, T. F. Synthesis of Graphene Aerogel with High Electrical Conductivity. *J. Am. Chem. Soc.* **2010**, *132* (40), 14067−14069.

44. Splendiani, A.; Sun, L.; Zhang, Y.; Li, T.; Kim, J.; Chim, C.-Y.; Galli, G.; Wang, F. Emerging Photoluminescence in Monolayer MoS2. *Nano Lett.* **2010**, *10* (4), 1271−1275.

45. Yu, W. J.; Liu, Y.; Zhou, H.; Yin, A.; Li, Z.; Huang, Y.; Duan, X. Highly Efficient Gate-Tunable Photocurrent Generation in Vertical Heterostructures of Layered Materials. *Nat. Nanotechnol.* **2013,** *8* (12), 952−958.

46. Roy, K.; Padmanabhan, M.; Goswami, S.; Sai, T. P.; Ramalingam, G.; Raghavan, S.; Ghosh, A. Graphene-MoS₂ Hybrid Structures for Multifunctional Photoresponsive Memory Devices. *Nat. Nano* **2013,** *8* (11), 826−830.

47. Zeng, S.; Hu, S.; Xia, J.; Anderson, T.; Dinh, X.-Q.; Meng, X.-M.; Coquet, P.; Yong, K.-T. Graphene−MoS₂ Hybrid Nanostructures Enhanced Surface Plasmon Resonance Biosensors. *Sensors Actuat. B Chem.* **2015,** *207* (Part A), 801−810.

48. Nitta, N.; Wu, F.; Lee, J. T.; Yushin, G. Li-ion Battery Materials: Present and Future. *Mater. Today* **2015,** *18* (5), 252−264.

49. Ratha, S.; Rout, C. S. Supercapacitor Electrodes Based on Layered Tungsten Disulfide-Reduced Graphene Oxide Hybrids Synthesized by a Facile Hydrothermal Method. *ACS Appl. Mater. Interfaces* **2013,** *5* (21), 11427−11433.

50. Rout, C. S.; Joshi, P. D.; Kashid, R. V.; Joag, D. S.; More, M. A.; Simbeck, A. J.; Washington, M.; Nayak, S. K.; Late, D. J. Superior Field Emission Properties of Layered WS₂-RGO Nanocomposites. *Sci. Rep.* **2013,** *3,* 3282.

51. Zhou, H.; Yu, F.; Sun, J.; He, R.; Wang, Y.; Guo, C. F.; Wang, F.; Lan, Y.; Ren, Z.; Chen, S. Highly Active and Durable Self-Standing WS₂/Graphene Hybrid Catalysts for the Hydrogen Evolution Reaction. *J. Mater. Chem. A* **2016,** *4* (24), 9472−9476.

52. Georgiou, T.; Jalil, R.; Belle, B. D.; Britnell, L.; Gorbachev, R. V.; Morozov, S. V.; Kim, Y.-J.; Gholinia, A.; Haigh, S. J.; Makarovsky, O., et al. Vertical Field-Effect Transistor Based on Graphene-WS2 Heterostructures for Flexible and Transparent Electronics. *Nat. Nano* **2013,** *8* (2), 100−103.

53. Huo, N.; Wei, Z.; Meng, X.; Kang, J.; Wu, F.; Li, S.-S.; Wei, S.-H.; Li, J. Interlayer Coupling and Optoelectronic Properties of Ultrathin Two-Dimensional Heterostructures Based on Graphene, MoS₂ and WS₂. *J. Mater. Chem. C* **2015,** *3* (21), 5467−5473.

54. Azizi, A.; Eichfeld, S.; Geschwind, G.; Zhang, K.; Jiang, B.; Mukherjee, D.; Hossain, L.; Piasecki, A. F.; Kabius, B.; Robinson, J. A. Freestanding Van Der Waals Heterostructures of Graphene and Transition Metal Dichalcogenides. *ACS Nano* **2015,** *9* (5), 4882−4890.

55. Kim, K.; Larentis, S.; Fallahazad, B.; Lee, K.; Xue, J.; Dillen, D. C.; Corbet, C. M.; Tutuc, E. Band Alignment in WSe₂−Graphene Heterostructures. *ACS Nano* **2015,** *9* (4), 4527−4532.

56. Shim, J.; Kim, H. S.; Shim, Y. S.; Kang, D. H.; Park, H. Y.; Lee, J.; Jeon, J.; Jung, S. J.; Song, Y. J.; Jung, W. S. Extremely Large Gate Modulation in Vertical Graphene/WSe₂ Heterojunction Barristor Based on a Novel Transport Mechanism. *Adv. Mater.* **2016,** *28* (26), 5293−5299.

57. Shim, G. W.; Yoo, K.; Seo, S.-B.; Shin, J.; Jung, D. Y.; Kang, I.-S.; Ahn, C. W.; Cho, B. J.; Choi, S.-Y. Large-Area Single-Layer MoSe2 and Its Van Der Waals Heterostructures. *ACS Nano* **2014,** *8* (7), 6655−6662.

58. Sata, Y.; Moriya, R.; Morikawa, S.; Yabuki, N.; Masubuchi, S.; Machida, T. Electric Field Modulation of Schottky Barrier Height in Graphene/MoSe₂ Van Der Waals Heterointerface. *Appl. Phys. Lett.* **2015,** *107* (2), 023109.

59. Kuiri, M.; Chakraborty, B.; Paul, A.; Das, S.; Sood, A.; Das, A. Enhancing Photoresponsivity Using MoTe₂-Graphene Vertical Heterostructures. *Appl. Phys. Lett.* **2016,** *108* (6), 063506.

60. Dang, W.; Peng, H.; Li, H.; Wang, P.; Liu, Z. Epitaxial Heterostructures of Ultrathin Topological Insulator Nanoplate and Graphene. *Nano Lett.* **2010,** *10* (8), 2870−2876.

61. Novoselov, K.; Jiang, D.; Schedin, F.; Booth, T.; Khotkevich, V.; Morozov, S.; Geim, A. Two-Dimensional Atomic Crystals. *Proc. Natl. Acad. Sci. USA* **2005,** *102* (30), 10451−10453.

62. Fu, L.; Sun, Y.; Wu, N.; Mendes, R. G.; Chen, L.; Xu, Z.; Zhang, T.; Rümmeli, M. H.; Rellinghaus, B.; Pohl, D. Direct Growth of MoS2/h-BN Heterostructures via a Sulfide-Resistant Alloy. *ACS Nano* **2016,** *10* (2), 2063−2070.

63. Antonelou, A.; Hoffman, T.; Edgar, J. H.; Yannopoulos, S. N. MoS2/h-BN Heterostructures: Controlling MoS_2 Crystal Morphology by Chemical Vapor Deposition. *J. Mater. Sci.* **2017,** *7*, 1−11.

64. Ye, L.; Li, H.; Chen, Z.; Xu, J. Near-Infrared Photodetector Based on MoS2/Black Phosphorus Heterojunction. *ACS Photon.* **2016,** *3* (4), 692−699.

65. Amin, B.; Kaloni, T. P.; Schreckenbach, G.; Freund, M. S. Materials Properties of Out-of-Plane Heterostructures of MoS_2-WSe_2 and WS_2-$MoSe_2$. *Appl. Phys. Lett.* **2016,** *108* (6), 063105.

66. Gong, Y.; Lin, J.; Wang, X.; Shi, G.; Lei, S.; Lin, Z.; Zou, X.; Ye, G.; Vajtai, R.; Yakobson, B. I. Vertical and In-Plane Heterostructures From WS_2/MoS_2 Monolayers. *Nat. Mater.* **2014,** *13* (12), 1135−1142.

67. Zhang, X.; Meng, F.; Christianson, J. R.; Arroyo-Torres, C.; Lukowski, M. A.; Liang, D.; Schmidt, J. R.; Jin, S. Vertical Heterostructures of Layered Metal Chalcogenides by Van Der Waals Epitaxy. *Nano Lett.* **2014,** *14* (6), 3047−3054.

68. Gong, Y.; Lei, S.; Ye, G.; Li, B.; He, Y.; Keyshar, K.; Zhang, X.; Wang, Q.; Lou, J.; Liu, Z. Two-Step Growth Of Two-Dimensional WSe_2/$MoSe_2$ Heterostructures. *Nano Lett.* **2015,** *15* (9), 6135−6141.

69. Ghuge, A.; Shirode, A.; Kadam, V. Graphene: A Comprehensive Review. *Curr. Drug Targets* **2016**. Available from: http://dx.doi.org/10.2174/13894501186661702271 53940.

70. Chhowalla, M.; Shin, H. S.; Eda, G.; Li, L.-J.; Loh, K. P.; Zhang, H. The Chemistry of Two-Dimensional Layered Transition Metal Dichalcogenide Nanosheets. *Nat. Chem.* **2013,** *5* (4), 263−275.

Graphene hybrids with carbon allotropes

6

CHAPTER OUTLINE

6.1 Summary 123
6.2 Introduction 124
6.3 Graphene/CNT Hybrids 125
    6.3.1 Covalent CNT/Graphene Hybrids 125
    6.3.2 Noncovalent CNT/Graphene Hybrids 127
6.4 Graphene/Fullerene Hybrids 129
6.5 Organic Chemistry of Graphene 129
6.6 Conclusions 134
References 135

6.1 SUMMARY

Second-generation graphene nanohybrids, along with other carbon allotropes, such as carbon nanotubes (CNTs) and fullerenes, are discussed in this chapter. Functional materials with CNTs and fullerenes are created by using covalent and noncovalent chemistry of carbon surfaces. The tethered three-dimensional architectures of these components display synergetic properties useful for energy storage, catalysis, sensing, and biology. The design, synthesis, and applications of graphene/CNT hybrids are discussed initially. The properties of graphene/CNT hybrid materials are primarily controlled by the orientation of CNTs in the graphene matrix. Vertically aligned CNT hybrids show superior properties in energy storage and catalysis due to the larger interlayer separation of the graphene layers (100 nm−20 μm) in the composite. Next, we will discuss the assembly of fullerenes on graphene sheets, where electronic coupling between electron-accepting fullerenes and electron-rich graphene sheets is observed. Graphene/C_{60} donor−acceptor systems are utilized in solar cell devices with about 4% photo-conversion efficiency using an all-carbon material, which is a fascinating design. We conclude our discussions with organic chemistry of graphene where highly reactive intermediates are used for the ring activation of graphene. This route opens a new door to the functionalization of graphene without having to employ harsh oxidation conditions. However, scale-up and improved efficiencies of these reactions are anticipated to happen in the near future. Collaboration between organic and material chemists might catalyze the development of a more

Introduction to Graphene. DOI: http://dx.doi.org/10.1016/B978-0-12-813182-4.00006-4

attractive chemistry of graphene in the future. Advances in graphene chemistry could be used to tune the bulk properties of graphene, such as its optoelectronic nature, with a molecular-level precision.

6.2 INTRODUCTION

After the advent of fullerenes in 1985, attention to nanocarbon allotropes escalated to explore their high surface area per mass, mechanical, and electronic properties.[1] The discovery of CNTs and graphene further catalyzed this field over three decades, from which stemmed the invention of functional materials with diverse applications of carbon origin. Considering the atomic structure, graphene, CNTs, and fullerenes are composed of sp^2 hybridized carbon with alternate single and double bonds, where π-electron conjugation is extended to nanoscale dimensions. Graphene is flat in the freestanding form, whereas fullerenes and CNTs have curved surfaces. Hybrid structures of graphene with its allotropes are envisioned as all-carbon functional materials which are applicable for electronics, catalysis, sensing, and medicine.[2] These hybrids that contain minimum levels of cross-linked nanocarbonaceous materials, or none at all, display electronic coupling between each other, allowing the synergetic effects of individual components to be explored. Covalent and noncovalent bonding between the components is expected to stabilize these structures in the assemblies.

Noncovalent interactions between the nanocarbon allotropes are dominated by hydrophobic, $\pi-\pi$ stacking and van der Waals interactions, in their pristine form. Upon modification of the carbon surface by oxidation or by other surface functionalizations, coulombic interactions and hydrogen bonding could also be introduced at the component interfaces. A combination of these forces is used to control the formation of the desired nanoarchitecture, since uncontrolled interactions lead to random aggregation of the components. Particularly, the face-to-face packing of graphene favors restacking of the layers. Hence, stabilizing the interface of graphene with the curved surfaces of CNTs and fullerenes requires a carefully designed synthesis method. Some examples of these strategies and synthetic protocols are discussed in the first two sections of this chapter. The final section of the chapter discusses the organic chemistry of pristine graphene.

Activation of aromatic rings in graphene by controlled chemical reactions leads to modulation of its optoelectronic properties.[3] Traditional organic chemistry principles and surface chemistry approaches are readily adopted in developing graphene chemistry. Inspired from the organic chemistry of benzene and polyaromatic hydrocarbons, cycloaddition reactions, diazonium chemistry, and radical additions are executed for benzenoid-like graphene structure.[3-6] A major challenge in this field is the characterization of the final product, where organic chemistry of small molecules relies on nuclear magnetic resonance (NMR) techniques. Graphene sheets are not suitable for solution-phase NMR spectroscopy due to their poor "solubility" (dispersibility for a colloid), poor yield, and large molar mass. However, surface characterization techniques such as IR, XPS, and Raman spectroscopy can address some of these problems, when the surface has been significantly modified.[4] Changes in electronic conductivity or carrier mobility, before and after functionalization, can be monitored, but may not be sensitive to minor modifications.[5] Some examples of the organic chemical reactions of graphene and how to better monitor these processes will be discussed in the final section of the chapter.

6.3 GRAPHENE/CNT HYBRIDS

CNT chemistry evolved in the 1990s, far before the discovery of graphene, and the findings in CNT research are broadly adopted to graphene. Using this new knowledge, CNT/graphene hybrid materials with high surface areas are designed for various applications. These hybrids are classified, based on the interaction at the interface, as covalent and noncovalent hybrid structures.

6.3.1 COVALENT CNT/GRAPHENE HYBRIDS

Chemically bonded three-dimensional structures of graphene and CNTs can be prepared by two major routes. First, both components are grown in situ or sequentially using chemical vapor deposition (CVD) of carbon sources on specific catalysts. Second, the covalent chemistry of surface groups on graphene (or its oxides) and CNT (derivatives) are extended to tailor these structures together (Table 6.1). The benefit of CVD-grown nanostructures is the complete conjugation of the π electrons in the network, which reduces the contact resistance.[7] For in situ growth methods, bifunctional catalysts (usually Al/Fe-based systems) are used for graphene and CNT.[7] Growth of CNT on top of an existing layer of graphene or graphene oxide (GO) after the deposition of the catalyst is an alternative route for controlled growth of CNT. The alignment of CNTs in the graphene matrix alters the properties of the hybrid.[1] Hybrids with vertically aligned CNT (also called pillared structures)[17] have high surface areas due to the larger separation of graphene sheets, when compared to horizontally aligned derivatives (Fig. 6.1). Randomly oriented graphene−CNT hybrids result when GO and functionalized CNT are linked by chemical crosslinking (Table 6.1). However, such composites can be prepared in large quantities compared to the CVD-grown hybrids.

The morphology of these hybrids is usually characterized by scanning electron microscopy images. The sandwiched CNT/graphene structures are visible in scanning electron microscopy (SEM) micrographs and the dimensions of the composites can be obtained (Fig. 6.2). The height of CNT pillars can be controlled from 100 nm to 20 μm, by controlling growth time and/or precursor concentrations, when a CVD method is employed.[16,17] The vertically aligned graphene−CNT

Table 6.1 Synthesis and Application of Covalently Linked Graphene—CNT Hybrids

System	Method	Application
Pristine SWCNT—graphene[8]	• CVD • In situ growth	Conductive/transparent films
Graphene—CNT pillars[17]	• CVD • In situ growth	NA
GO—CNT pillars[16]	• CVD • In situ growth of CNT on thermally annealed GO	Supercapacitor
rGO—CNT[9]	• Hummers method and CVD • In situ growth of CNT on rGO	Mechanically durable Conductive/transparent films
Graphene—CNT[10]	• CVD on silica nanoparticle (SiNP) • Simultaneous growth of CNT and graphene on SiNP	Field effect transistor
Graphene—SWCNT[11]	• CVD growth of graphene on top of CNT	Photodetectors (proposed)
Graphene—CNT7	• In situ CVD growth using Fe_3O_4/AlOx nanocatalyst	Li-ion capacitors
Graphene—CNT[12]	• Sequential CVD growth	Energy storage (proposed)
Graphene—aligned CNT[13]	• CVD using bifunctional catalyst	Li-S battery
GO—MWCNT[14]	• Amide coupling using carbodiimide	Water purification
GO—MWCNT[15]	• Amide coupling of GO to p-phenylenediamine • p-Phenylenediamine-modified GO was linked to CNT by diazonium coupling	Supercapacitor and dye removal

hybrids are especially important in energy storage or catalysis applications, due to their surface area. About a threefold increase in surface area is seen after expansion with in situ growth of CNT on GO (~ 200 m^2/g), which improved the specific capacitance of GO from ~ 140 F/g to ~ 390 F/g.[16] The increase in capacitance of these hybrids is attributed to the effective separation of graphene by CNT spacers. The increase in surface area of the composite and the presence of chemical functional groups on the surfaces of the component phase are utilized to fabricate electronic devices including supercapacitors and Li-ion batteries. Some examples of such attempts are presented in Table 6.1. Next, we examine the synthesis and properties of graphene/CNT noncovalent assemblies.

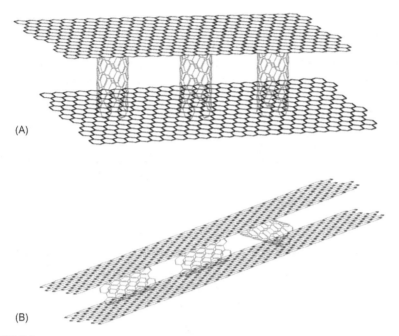

FIGURE 6.1

(A) Vertically aligned CNT/graphene nanostructures (also called pillared CNT—graphene structure). (B) Horizontally aligned CNT/graphene structures.

FIGURE 6.2

SEM image of graphene—CNT pillared structure prepared by chemical vapor deposition method (CVD) (© John Wiley and Sons).[17]

6.3.2 NONCOVALENT CNT/GRAPHENE HYBRIDS

Hierarchal structures of graphene/CNT are assembled by favorable hydrophobic interactions, $\pi-\pi$ stacking, electrostatic interactions, hydrogen bonding, and van der Waals interactions between the component phases.[1] Complementary binding groups are tethered to graphene and/or CNT surfaces to ensure attractive interactions between the two components. Hence, most of these composites start with GO as the graphene precursor and pristine, oxidized or amino-modified CNT as the CNT component (Table 6.2). When the CNT is unmodified, hydrophobic interactions and $\pi-\pi$ stacking operate at the interface. Strong electrostatic attraction and H-bonding were implemented by modifying CNT with positively charged amine groups (at neutral pH).[1] Aqueous stability of modified CNTs is another advantage for mass production of devices from these carbon materials. Topological studies suggested a randomly organized structure, where graphene and CNTs are entangled together in the hybrid. However, layer-by-layer assembly of oppositely charged GO and amino-CNT was also used to make transparent

Table 6.2 Synthesis And application of Noncovalently Linked Graphene/CNT Hybrids

Type of CNT	Type of Graphene	Major Interaction	Application
Pristine: grown from hydroquinone by CVD[18]	GO by Hummers method	Hydrophobic and van der Waals interactions	Li-ion battery
• Pristine SWCNT • Boron-doped SWCNT • Nitrogen-doped SWCNT[19]	GO	Electrostatic layer-by-layer assembly using polycationic surfactants[20]	Li-ion battery
Pristine MWCNT[21]	rGO	$\pi-\pi$ interaction through a pyrene linker	Biosensor for dopamine
Pristine MWCNT[22]	GO	Hydrophobic and van der Waals interactions after freeze dying	Microbial biofuel cell
Amino-CNT[23]	GO by Hummers method	Hydrogen bonding and electrostatic	Biofuel cell
Amino-MWCNT[24]	GO by Hummers method	Hydrogen bonding and electrostatic	Conductive/ transparent electrode
Amino-CNT[25]	GO by Hummers method	Hydrogen bonding and electrostatic	Energy storage
Amino-MWCNT[26]	rGO	Hydrogen bonding and electrostatic	Conductive/ transparent thin films
Oxidized-CNT[27]	GO/poly (etheleneimine)	Hydrogen bonding and electrostatic	Supercapacitor
Oxidized-CNT[28]	rGO	NA	Solar cell

conductive films.[19] These hybrid structures find applications in batteries, biosensors, biofuel cells, and supercapacitors as electrode materials, which are superior to those made from their individual components. A review of some of these hybrids is given in Table 6.2.

6.4 GRAPHENE/FULLERENE HYBRIDS

Graphene fullerene assemblies are commonly fabricated by solution-phase methods.[29] Since fullerenes are employed as an active component in organic solar cells, their graphene composites find applications in photo energy conversion. The layered structure of graphene allows for uniform fullerene assembly, without particle aggregation, by covalent or noncovalent grafting techniques (Table 6.3). Noncovalent assemblies usually explore the $\pi-\pi$ interaction between a spacer group (e.g., pyrene) attached to the fullerene and graphene π system.[30] Nucleophilic addition to C_{60} was explored in these composites by using organolithium reagents, followed by addition of an electrophile to the carbanion intermediate of C_{60}.[31] Ball milling of graphite with C_{60} in the presence of LiOH enabled the simultaneous exfoliation and grafting of C_{60}, which is a promising route for mass production.[32] Some different approaches used to synthesize these systems are tabulated (Table 6.3). The structures of fullerenes on graphene are examined by Raman spectroscopy and transmission electron microscopy (TEM) images, as well as by other methods. TEM images clearly show the fullerenes are separated on graphene sheets (Fig. 6.3). Electronic coupling between electron-deficient C_{60} and electron-rich graphene is also observed at this interface, which can be used for photo-detection devices

Supercapacitors fabricated using these hybrids exploit fullerenes that can act as spacer molecules between graphene layers (sandwiched structure). This can increase the surface area and specific capacitance of the active material (upto 350 F/g).[33] The electron-rich property of graphene and the electron-accepting nature of C_{60} were explored as a donor−acceptor system for photo-induced electron transfer reactions.[36,37] The photo response of the hybrids is explored in in vitro photodynamic therapy as well.[34] The promising application of fullerenes in solar cells is also extended for its graphene hybrid, where up to 4% photo-conversion efficiencies are attained.[30] Some properties and applications of these hybrid materials are given in Table 6.3.

6.5 ORGANIC CHEMISTRY OF GRAPHENE

Organic chemistry of pristine graphene is an emerging area where functionalization of the graphene surface with small molecules is used to tune the optoelectronic properties of graphene. Although Hummers method uses the oxidation of graphene layers, it produces a severely oxygenated surface, and thus, the electronic

Table 6.3 Synthesis and Application of Noncovalently Linked C_{60}/Graphene Hybrids

Hybrid Material	Synthesis	Properties and Applications
C_{60}/GO[29]	Solution-phase method using Li-modified C_{60} and GO	• C_{60} acts as a spacer for GO sheets • Supercapacitor with specific capacitance $\sim$135 F/g
Cationic C_{60}/GO[33]	• N,N-dimethylpyrrolidinium iodide and N-methylpyrrolidine-modified C_{60} • Dispersed in ethanol/water mixture	Supercapacitor with specific capacitance $\sim$350 F/g
C_{60}/GO[34]	Nucleophilic addition of GO-PEG-NH_2 to C_{60}	Photodynamic therapy
C_{60}/GO[35]	Fullerenol (C_{60}-$(OH)_n$) was grafted to GO by esterification reaction	NA
C_{60}/GO[36]	Phenyl butyric acid methyl ester-modified C_{60} was covalently attached to GO via an esterification reaction	Ultrafast electron transfer from GO to C_{60}
C_{60}/GO[37]	GO and C_{60} were tethered by alkyne-azide cycloaddition reaction	Photo-induced electron transfer from GO to C_{60}
C_{60}/rGO[38]	rGO and C_{60} are grafted by lithiation reaction	Bulk heterojunction solar cell with efficiency $\sim$1.22%
C_{60}/rGO[39]	Polydopamine was used as a crosslinker for rGO and C_{60}	Protecting agent against nitrogen monoxide-induced cytotoxicity in rat pheochromocytoma cells
C_{60}/rGO[30]	Pyrene butyric acid-modified C_{61} was assembled to rGO by π–π interaction	Bulk heterojunction solar cell with photo-conversion efficiency $\sim$4%
C_{60}/rGO[40]	Zinc porphyrin-modified C_{60} was noncovalently attached to electrochemically reduced GO	Electrochemical sensor for H_2O_2 (detection limit: 0.27 μM)
C_{60}/graphene[41]	• Graphene by CVD • Thermal deposition of C_{60}	Long-lived photoconductivity (100 ps)
C_{60}/graphene[32]	Ball milling of graphite and C_{60} with LiOH	Catalyst for oxygen reduction reaction

and mechanical properties of the resulting GO are significantly different from those of graphene. The rich chemistry of GO functional groups is efficiently explored for functional materials in diverse fields including chemistry, physics, and biology, as well as materials science.[42] The poor dispersive nature of unoxidized graphene in solvents was a challenge until successful liquid-phase exfoliation techniques were introduced. The reactions of CVD-grown graphene and exfoliated graphene[43] with small molecules are still being explored, as discussed below.

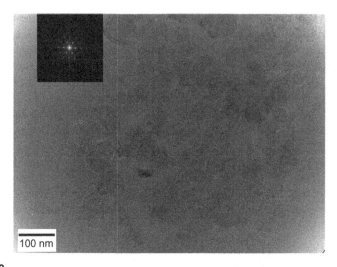

FIGURE 6.3

TEM image of C_{60} grafted graphene sheet. Fullerenol was covalently grafted to graphene oxide by an ester bond (scale bar is 100 nm). Inset shows the selected area electron diffraction (SAED) pattern corresponding to few-layer graphene (© Royal Society of Chemistry.[35]

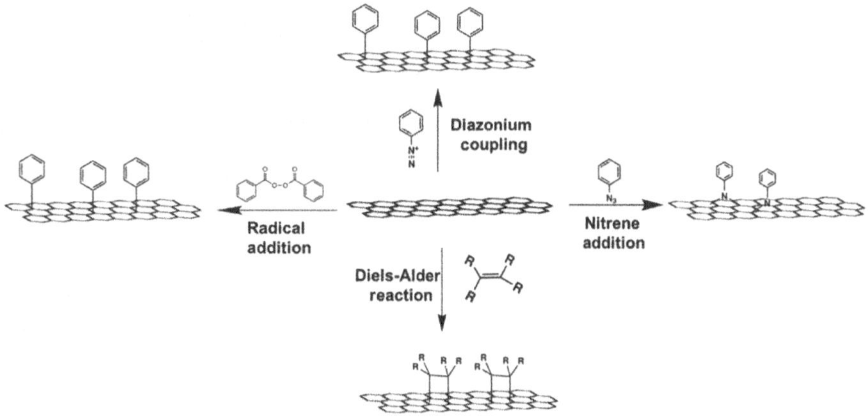

FIGURE 6.4

Some organic reactions of pristine graphene.

Benzenoid structures in the graphene basal plane are chemically inert and not available for common reactions (Fig. 6.4). To activate the rings, highly reactive intermediates were used. Organic radicals, e.g., were explored to be used for graphene chemistry. Diazonium chemistry is particularly interesting because it can be prepared from the corresponding amines and the radicals are generated by thermal or electrochemical activation.[44] Commercial availability of 4-nitrophenyl diazonium tetrafluoroborate is also an advantage for large-scale production of the diazonium-modified derivatives.[49] Diels—Alder reactions of graphene "diene" and dienophiles are explored in the solution phase.[53] Photo-initiated radical polymerization of styrene is another example of how graphene chemistry is coupled with functional polymers.[52] Other reactions involving highly reactive intermediates include nitrenes, peroxides, and arynes, and a few examples are given in Table 6.4.

The versatility of Raman spectroscopy in graphene research has proved once again to be very useful in the characterization of small-molecule modified graphene (Table. 6.4).[3] The evolution of the D peak due to the insertion of small molecules into the graphene plane effectively represents the modification in most

Table 6.4 List of Organic Reactions of Pristine Graphene and Characterization Techniques

Graphene Type	Reactive Molecule	Key Reaction	Major Characterization Techniques
CVD[44]	4-Docosyloxy-benzenediazonium tetrafluoroborate	Electrochemically activated aryl diazonium coupling to graphene	Scanning tunneling microscopy (STM), AFM, XPS, cyclic voltammetry, and Raman spectroscopy
CVD[45]	4-Nitrophenyl diazonium tetrafluoroborate	Aryl diazonium coupling aided by silica nanoparticle	MicroRaman mapping, XPS, and AFM
CVD[46]	4-Nitrophenyl diazonium tetrafluoroborate	Aryl diazonium coupling	Spatial Raman mapping, AFM, and Raman spectroscopy
Exfoliated using Scotch tape[47]	4-Nitrophenyl diazonium tetrafluoroborate	Aryl diazonium coupling	Raman spectroscopy, angle-resolved photoemission spectroscopy, and AFM
Exfoliated using Scotch tape[48]	4-Bromobenzene diazonium tetrafluoroborate	Aryl diazonium coupling	Raman spectroscopy, scanning photoelectron microscopy, spatially resolved XPS, and AFM
Electrochemically exfoliated[49]	4-Nitrophenyl diazonium tetrafluoroborate	In situ electrochemical exfoliation and diazonium coupling	Raman spectroscopy, SEM, TEM, absorption spectroscopy, AFM, and XPS

(Continued)

Table 6.4 List of Organic Reactions of Pristine Graphene and Characterization Techniques *Continued*

Graphene Type	Reactive Molecule	Key Reaction	Major Characterization Techniques
Graphite[50]	L-phenylalanine	Friedel–Crafts acylation reaction in PPA/P$_2$O$_5$	Raman spectroscopy, FTIR, thermogravimetric analysis, XRD, SEM, and TEM
Exfoliated in *o*-dichlorobenzene by ultrasonication[51]	Perfluorophenylazide derivatives	Photo/thermal addition of singlet perfluorophenyl nitrene to C = C	Raman spectroscopy, FTIR, XPS, and TEM
CVD[52]	Styrene	UV-induced polymerization of styrene on graphene	Confocal Raman spectroscopy, FTIR, and AFM
Exfoliated in *o*-dichlorobenzene by ultrasonication[53]	Dienophile tetracyanoethylene, maleic anhydride, 9-methylanthracene and 2,3-dimethoxy-1,3-butadiene	Diels–Alder reaction	Raman spectroscopy and FTIR
Thermal evaporation of Si from SiC[54]	3,5-Bis(trifluoromethyl) phenyl-substituted maleimide derivatives	Diels–Alder reaction	Scanning tunneling microscopy, angle-resolved photoemission spectroscopy, XPS, and Raman spectroscopy
CVD[55]	Dihydronaphthalene and indene	Diels–Alder reaction	Raman spectroscopy, AFM, XPS, and scanning kelvin probe microscopy
Exfoliated using Scotch tape[56]	Benzoyl peroxide	Photochemical radical addition	Raman spectroscopy, optical imaging, and AFM
Epitaxial graphene[57]	Azidotrimethylsilane	Nitrene radical addition (thermal)	High-resolution photoemission spectroscopy
Arc discharge[58]	2-Triflatophenyl silane benzyne precursors	Aryne addition	Raman spectroscopy, AFM, XPS and HR-TEM, thermogravimetric analysis

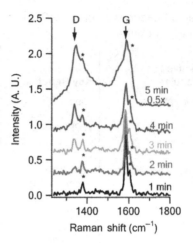

FIGURE 6.5

Evolution of D band of single-layer graphene over time upon reaction with benzoyl peroxide. *Indicate Raman peaks caused by the toluene (5 mM) in which the reaction was carried out. (©American Chemical Society).[56]

cases (Fig. 6.5).[55] High-resolution imaging using electron and X-ray microscopes can also be used to assess the atomic-level defects or vacancies in the modified graphene.[4] Alternatively, scanning probe microscopy and AFMs are used to image graphene on different surfaces to prove basal plane modifications.[4] We recommend the reader refer to the original articles for better insight into these characterization techniques, as they vary with the synthesis method of graphene and the modified molecule.

6.6 CONCLUSIONS

Connecting zero-, one- and two-dimensional nanocarbon allotropes by covalent and noncovalent interactions often results in all-carbon hybrid materials with synergetic properties. The beauty of nanotechnology is underlined in these 3-D architectures where not only the size, but also the geometry, of the materials at nanoscale affects the properties. The versatility of carbon allotropes at nanoscale is readily harnessed in these hybrid materials to explore their uses in energy storage, photodetectors, biosensors, drug delivery, and so on, by controlling the three-dimensional geometry of the components in the all-carbon composites. CVD techniques are useful to amalgamate graphene with CNT where the catalysts guide the formation of graphene or CNT from a single carbon source.[15] Such chemically bonded graphene−CNT hybrids show no contact resistance when compared to graphene/CNT covalent structures. Orientation of CNT in the graphene matrix can also be controlled by CVD methods. Vertical orientation alignments display maximum surface area, which is advantageous for energy storage and catalysis applications. Since fullerenes are not grown using CVD, most of the graphene/C_{60} analogues are noncovalently attached or connected via linkers (Table 6.3). The electron-accepting nature of fullerenes and electron-donating nature of graphene were utilized in both covalent and noncovalent graphene/C_{60} hybrids.[41]

Organic chemistry of graphene basal plane systems is being rapidly investigated because of the poor reactivity of benzenoid rings in graphene. However, strongly reactive (thus short-lived) organic intermediates activate the graphene plane and the products can be characterized using various spectroscopic and microscopic techniques (Table 6.4). Applications of such systems are still to be explored in light of the potential high impact of these reactions. Specifically, a major thrust toward bioconjugation is anticipated. Complete or near-complete reduction of the aromatic system in graphene would lead to the formation of highly strained graphane[59], a highly awaited member of this family of these materials in need of mass production. Theoretical studies predict a rapid conversion of graphane to graphene by releasing hydrogen atoms, which could be useful for hydrogen storage applications.[60] Organic chemistry of graphene and the formation of its hybrids can be taken as the "Graphene 2.0" era, where the hybrid materials display multitasking properties, which could facilitate another giant step for real-life applications.

REFERENCES

1. Georgakilas, V.; Perman, J. A.; Tucek, J.; Zboril, R. Broad Family of Carbon Nanoallotropes: Classification, Chemistry, and Applications of Fullerenes, Carbon Dots, Nanotubes, Graphene, Nanodiamonds, and Combined Superstructures. *Chem. Rev.* **2015,** *115* (11), 4744−4822.
2. Wang, X.; Sun, G.; Chen, P. Three-Dimensional Porous Architectures of Carbon Nanotubes and Graphene Sheets for Energy Applications. *Front. Energy Res.* **2014,** *2*, 33.
3. Criado, A.; Melchionna, M.; Marchesan, S.; Prato, M. The Covalent Functionalization of Graphene on Substrates. *Angew. Chem. Int. Ed.* **2015,** *54* (37), 10734−10750.
4. Koehler, F. M.; Stark, W. J. Organic Synthesis on Graphene. *Acc. Chem. Res.* **2012,** *46* (10), 2297−2306.
5. Quintana, M.; Vazquez, E.; Prato, M. Organic Functionalization of Graphene in Dispersions. *Acc. Chem. Res.* **2012,** *46* (1), 138−148.
6. Park, J.; Yan, M. Covalent Functionalization of Graphene With Reactive Intermediates. *Acc. Chem. Res.* **2012,** *46* (1), 181−189.
7. Salvatierra, R. V.; Zakhidov, D.; Sha, J.; Kim, N. D.; Lee, S.-K.; Raji, A.-R. O.; Zhao, N.; Tour, J. M. Graphene Carbon Nanotube Carpets Grown Using Binary Catalysts for High-Performance Lithium-Ion Capacitors. *ACS Nano* **2017,** *11* (3), 2724−2733.
8. Maarouf, A. A.; Kasry, A.; Chandra, B.; Martyna, G. J. A Graphene−Carbon Nanotube Hybrid Material for Photovoltaic Applications. *Carbon. N. Y.* **2016,** *102*, 74−80.
9. Lee, D. H.; Kim, J. E.; Han, T. H.; Hwang, J. W.; Jeon, S.; Choi, S. Y.; Hong, S. H.; Lee, W. J.; Ruoff, R. S.; Kim, S. O. Versatile Carbon Hybrid Films Composed of

Vertical Carbon Nanotubes Grown on Mechanically Compliant Graphene Films. *Adv. Mater.* **2010,** *22* (11), 1247−1252.

10. Dong, X.; Li, B.; Wei, A.; Cao, X.; Chan-Park, M. B.; Zhang, H.; Li, L.-J.; Huang, W.; Chen, P. One-Step Growth of Graphene−Carbon Nanotube Hybrid Materials by Chemical Vapor Deposition. *Carbon. N. Y.* **2011,** *49* (9), 2944−2949.

11. Wang, R.; Hong, T.; Xu, Y.-Q. Ultrathin Single-Walled Carbon Nanotube Network Framed Graphene Hybrids. *ACS Appl. Mater. Interfaces* **2015,** *7* (9), 5233−5238.

12. Jiang, J.; Li, Y.; Gao, C.; Kim, N. D.; Fan, X.; Wang, G.; Peng, Z.; Hauge, R. H.; Tour, J. M. Growing Carbon Nanotubes From Both Sides of Graphene. *ACS Appl. Mater. Interfaces* **2016,** *8* (11), 7356−7362.

13. Tang, C.; Zhang, Q.; Zhao, M. Q.; Huang, J. Q.; Cheng, X. B.; Tian, G. L.; Peng, H. J.; Wei, F. Nitrogen-Doped Aligned Carbon Nanotube/Graphene Sandwiches: Facile Catalytic Growth on Bifunctional Natural Catalysts and Their Applications as Scaffolds for High-Rate Lithium-Sulfur Batteries. *Adv. Mater.* **2014,** *26* (35), 6100−6105.

14. Nellore, B. P. V.; Kanchanapally, R.; Pedraza, F.; Sinha, S. S.; Pramanik, A.; Hamme, A. T.; Arslan, Z.; Sardar, D.; Ray, P. C. Bio-Conjugated CNT-Bridged 3D Porous Graphene Oxide Membrane for Highly Efficient Disinfection of Pathogenic Bacteria and Removal of Toxic Metals From Water. *ACS Appl. Mater. Interfaces* **2015,** *7* (34), 19210.

15. Kotal, M.; Bhowmick, A. K. Multifunctional Hybrid Materials Based on Carbon Nanotube Chemically Bonded to Reduced Graphene Oxide. *J. Phys. Chem. C* **2013,** *117* (48), 25865−25875.

16. Fan, Z.; Yan, J.; Zhi, L.; Zhang, Q.; Wei, T.; Feng, J.; Zhang, M.; Qian, W.; Wei, F. A Three-Dimensional Carbon Nanotube/Graphene Sandwich and Its Application as Electrode in Supercapacitors. *Adv. Mater.* **2010,** *22* (33), 3723−3728.

17. Paul, R. K.; Ghazinejad, M.; Penchev, M.; Lin, J.; Ozkan, M.; Ozkan, C. S. Synthesis of a Pillared Graphene Nanostructure: A Counterpart of Three-Dimensional Carbon Architectures. *Small.* **2010,** *6* (20), 2309−2313.

18. Sridhar, V.; Lee, I.; Chun, H.-H.; Park, H. Hydroquinone as a Single Precursor for Concurrent Reduction and Growth of Carbon Nanotubes on Graphene Oxide. *RSC Adv.* **2015,** *5* (84), 68270−68275.

19. Tristán-López, F.; Morelos-Gómez, A.; Vega-Díaz, S. M.; García-Betancourt, M. L.; Perea-López, N.; Elías, A. L.; Muramatsu, H.; Cruz-Silva, R.; Tsuruoka, S.; Kim, Y. A. Large Area Films of Alternating Graphene−Carbon Nanotube Layers Processed in Water. *ACS Nano* **2013,** *7* (12), 10788−10798.

20. Paloniemi, H.; Ääritalo, T.; Laiho, T.; Liuke, H.; Kocharova, N.; Haapakka, K.; Terzi, F.; Seeber, R.; Lukkari, J. Water-Soluble Full-Length Single-Wall Carbon Nanotube Polyelectrolytes: Preparation and Characterization. *J. Phys. Chem. B* **2005,** *109* (18), 8634−8642.

21. Wang, Q.; Wang, S.; Shang, J.; Qiu, S.; Zhang, W.; Wu, X.; Li, J.; Chen, W.; Wang, X. Enhanced Electronic Communication and Electrochemical Sensitivity Benefiting from the Cooperation of Quadruple Hydrogen Bonding and π−π Interactions in Graphene/Multi-Walled Carbon Nanotube Hybrids. *ACS Appl. Mater. Interfaces* **2017,** *9* (7), 6255−6264.

22. Zou, L.; Qiao, Y.; Wu, X.-S.; Li, C. M. Tailoring Hierarchically Porous Graphene Architecture by Carbon Nanotube to Accelerate Extracellular Electron Transfer of Anodic Biofilm in Microbial Fuel Cells. *J. Power Sour.* **2016,** *328*, 143−150.

23. Navaee, A.; Salimi, A. Graphene-Supported Pyrene-Functionalized Amino-Carbon Nanotube: A Novel Hybrid Architecture of Laccase Immobilization as Effective Bioelectrocatalyst for Oxygen Reduction Reaction. *J. Mater. Chem. A* **2015,** *3* (14), 7623−7630.

24. Kim, Y.-K.; Min, D.-H. Durable Large-Area Thin Films of Graphene/Carbon Nanotube Double Layers as a Transparent Electrode. *Langmuir.* **2009,** *25* (19), 11302−11306.

25. Wu, C.; Huang, X.; Wu, X.; Xie, L.; Yang, K.; Jiang, P. Graphene Oxide-Encapsulated Carbon Nanotube Hybrids for High Dielectric Performance Nanocomposites with Enhanced Energy Storage Density. *Nanoscale* **2013,** *5* (9), 3847−3855.

26. Hong, T.-K.; Lee, D. W.; Choi, H. J.; Shin, H. S.; Kim, B.-S. Transparent, Flexible Conducting Hybrid Multilayer Thin Films of Multiwalled Carbon Nanotubes with Graphene Nanosheets. *Acs Nano* **2010,** *4* (7), 3861−3868.

27. Yu, D.; Dai, L. Self-Assembled Graphene/Carbon Nanotube Hybrid Films for Supercapacitors. *J. Phys. Chem. Lett.* **2009,** *1* (2), 467−470.

28. Tung, V. C.; Chen, L.-M.; Allen, M. J.; Wassei, J. K.; Nelson, K.; Kaner, R. B.; Yang, Y. Low-Temperature Solution Processing of Graphene − Carbon Nanotube Hybrid Materials for High-Performance Transparent Conductors. *Nano Lett.* **2009,** *9* (5), 1949−1955.

29. Ma, J.; Guo, Q.; Gao, H.-L.; Qin, X. Synthesis of C_{60}/Graphene Composite as Electrode in Supercapacitors. *Fuller. Nanotub. Carbon Nanostruct.* **2015,** *23* (6), 477−482.

30. Qu, S.; Li, M.; Xie, L.; Huang, X.; Yang, J.; Wang, N.; Yang, S. Noncovalent Functionalization of Graphene Attaching [6, 6]-phenyl-C61-Butyric Acid Methyl Ester (PCBM) and Application as Electron Extraction Layer of Polymer Solar Cells. *ACS Nano* **2013,** *7* (5), 4070−4081.

31. Hirsch, A.; Brettreich, M. *Nucleophilic Additions. Fullerenes;* Wiley-VCH Verlag GmbH & Co. KGaA, 2005, 73−99.

32. Guan, J.; Chen, X.; Wei, T.; Liu, F.; Wang, S.; Yang, Q.; Lu, Y.; Yang, S. Directly Bonded Hybrid of Graphene Nanoplatelets and Fullerene: Facile Solid-State Mechanochemical Synthesis and Application as Carbon-Based Electrocatalyst for Oxygen Reduction Reaction. *J. Mater. Chem. A* **2015,** *3* (8), 4139−4146.

33. Senthilkumar, K.; Prabakar, S. R.; Park, C.; Jeong, S.; Lah, M. S.; Pyo, M. Graphene Oxide Self-Assembled with a Cationic Fullerene for High Performance Pseudo-Capacitors. *J. Mater. Chem. A* **2016,** *4* (5), 1663−1670.

34. Hu, Z.; Li, J.; Huang, Y.; Chen, L.; Li, Z. Functionalized Graphene/C_{60} Nanohybrid for Targeting Photothermally Enhanced Photodynamic Therapy. *RSC Adv.* **2015,** *5* (1), 654−664.

35. Kumar, R.; Kumar, P.; Naqvi, S.; Gupta, N.; Saxena, N.; Gaur, J.; Maurya, J. K.; Chand, S. Stable Graphite Exfoliation by Fullerenol Intercalation via Aqueous Route. *New J. Chem.* **2014,** *38* (10), 4922−4930.

36. Kumar, R.; Khan, S.; Gupta, N.; Naqvi, S.; Gaurav, K.; Sharma, C.; Kumar, M.; Kumar, P.; Chand, S. Fullerene Grafted Graphene Oxide with Effective Charge Transfer Interactions. *Carbon. N. Y.* **2016,** *107*, 765−773.

37. Barrejón, M.; Vizuete, M.; Gómez-Escalonilla, M.; Fierro, J.; Berlanga, I.; Zamora, F.; Abellán, G.; Atienzar, P.; Nierengarten, J.-F.; García, H. A Photoresponsive Graphene Oxide−C 60 Conjugate. *Chem. Commun.* **2014,** *50* (65), 9053−9055.

38. Yu, D.; Park, K.; Durstock, M.; Dai, L. Fullerene-Grafted Graphene for Efficient Bulk Heterojunction Polymer Photovoltaic Devices. *J. Phys. Chem. Lett.* **2011,** *2* (10), 1113–1118.

39. Hu, Z.; Huang, Y.; Zhang, C.; Liu, L.; Li, J.; Wang, Y. Graphene–Polydopamine–C 60 Nanohybrid: An Efficient Protective Agent for NO-Induced Cytotoxicity in Rat Pheochromocytoma Cells. *J. Mater. Chem. B* **2014,** *2* (48), 8587–8597.

40. Fan, S.; Wei, T.; Zhang, J.; Zhang, N.; Chai, M.; Jin, X.; Wu, H. Zinc Porphyrin–Fullerene Derivative Noncovalently Functionalized Graphene Hybrid as Interfacial Material for Electrocatalytic Application. *Talanta.* **2016,** *160*, 713–720.

41. Jnawali, G.; Rao, Y.; Beck, J. H.; Petrone, N.; Kymissis, I.; Hone, J.; Heinz, T. F. Observation of Ground-and Excited-State Charge Transfer at the C_{60}/Graphene Interface. *ACS Nano* **2015,** *9* (7), 7175–7185.

42. Dreyer, D. R.; Park, S.; Bielawski, C. W.; Ruoff, R. S. The Chemistry of Graphene Oxide. *Chem. Soc. Rev.* **2010,** *39* (1), 228–240.

43. Economopoulos, S. P.; Tagmatarchis, N. Chemical Functionalization of Exfoliated Graphene. *Chem. A Eur. J.* **2013,** *19* (39), 12930–12936.

44. Xia, Z.; Leonardi, F.; Gobbi, M.; Liu, Y.; Bellani, V.; Liscio, A.; Kovtun, A.; Li, R.; Feng, X.; Orgiu, E. Electrochemical Functionalization of Graphene at the Nanoscale with Self-Assembling Diazonium Salts. *ACS Nano* **2016,** *10* (7), 7125–7134.

45. Wu, Q.; Wu, Y.; Hao, Y.; Geng, J.; Charlton, M.; Chen, S.; Ren, Y.; Ji, H.; Li, H.; Boukhvalov, D. W., et al. Selective Surface Functionalization at Regions of High Local Curvature in Graphene. *Chem. Commun.* **2013,** *49* (7), 677–679.

46. Wang, Q. H.; Jin, Z.; Kim, K. K.; Hilmer, A. J.; Paulus, G. L.; Shih, C.-J.; Ham, M.-H.; Sanchez-Yamagishi, J. D.; Watanabe, K.; Taniguchi, T. Understanding and Controlling the Substrate Effect on Graphene Electron-Transfer Chemistry via Reactivity Imprint Lithography. *Nat. Chem.* **2012,** *4* (9), 724–732.

47. Niyogi, S.; Bekyarova, E.; Itkis, M. E.; Zhang, H.; Shepperd, K.; Hicks, J.; Sprinkle, M.; Berger, C.; Lau, C. N.; Deheer, W. A. Spectroscopy of Covalently Functionalized Graphene. *Nano. Lett.* **2010,** *10* (10), 4061–4066.

48. Lim, H.; Lee, J. S.; Shin, H.-J.; Shin, H. S.; Choi, H. C. Spatially Resolved Spontaneous Reactivity of Diazonium Salt on Edge and Basal Plane of Graphene Without Surfactant and Its Doping Effect. *Langmuir.* **2010,** *26* (14), 12278–12284.

49. Ejigu, A.; Kinloch, I. A.; Dryfe, R. A. Single Stage Simultaneous Electrochemical Exfoliation and Functionalization of Graphene. *ACS Appl. Mater. Interfaces* **2016,** *28*, 6253–6261.

50. Abdolmaleki, A.; Mallkpour, S.; Borandeh, S. Structure, Morphology and Electronic Properties of L-Phenylalanine Edge-Functionalized Graphite Platelets Through Friedel–Crafts Acylation Reaction. *RSC Adv.* **2014,** *4* (104), 60052–60057.

51. Liu, L.-H.; Lerner, M. M.; Yan, M. Derivitization of Pristine Graphene with Well-Defined Chemical Functionalities. *Nano. Lett.* **2010,** *10* (9), 3754.

52. Steenackers, M.; Gigler, A. M.; Zhang, N.; Deubel, F.; Seifert, M.; Hess, L. H.; Lim, C. H. Y. X.; Loh, K. P.; Garrido, J. A.; Jordan, R. Polymer Brushes on Graphene. *J. Am. Chem. Soc.* **2011,** *133* (27), 10490–10498.

53. Sarkar, S.; Bekyarova, E.; Niyogi, S.; Haddon, R. C. Diels – Alder Chemistry of Graphite and Graphene: Graphene as Diene and Dienophile. *J. Am. Chem. Soc.* **2011,** *133* (10), 3324–3327.

54. Daukiya, L.; Mattioli, C.; Aubel, D.; Hajjar-Garreau, S.; Vonau, F.; Denys, E.; Reiter, G.; Fransson, J.; Perrin, E.; Bocquet, M.-L. Covalent Functionalization by Cycloaddition Reactions of Pristine Defect-Free Graphene. *ACS Nano* **2017,** *11* (1), 627−634.

55. Li, J.; Li, M.; Zhou, L.-L.; Lang, S.-Y.; Lu, H.-Y.; Wang, D.; Chen, C.-F.; Wan, L.-J. Click and Patterned Functionalization of Graphene by Diels−Alder Reaction. *J. Am. Chem. Soc.* **2016,** *138* (24), 7448−7451.

56. Liu, H.; Ryu, S.; Chen, Z.; Steigerwald, M. L.; Nuckolls, C.; Brus, L. E. Photochemical Reactivity of Graphene. *J. Am. Chem. Soc.* **2009,** *131* (47), 17099−17101.

57. Choi, J.; Kim, K.-j; Kim, B.; Lee, H.; Kim, S. Covalent Functionalization of Epitaxial Graphene by Azidotrimethylsilane. *J. Phys. Chem. C* **2009,** *113* (22), 9433−9435.

58. Zhong, X.; Jin, J.; Li, S.; Niu, Z.; Hu, W.; Li, R.; Ma, J. Aryne Cycloaddition: Highly Efficient Chemical Modification of Graphene. *Chem. Commun.* **2010,** *46* (39), 7340−7342.

59. Sahin, H.; Leenaerts, O.; Singh, S.; Peeters, F., GraphAne: from synthesis to applications. *arXiv preprint arXiv:1502.05804* **2015.**

60. Sofo, J. O.; Chaudhari, A. S.; Barber, G. D. Graphane: A Two-Dimensional Hydrocarbon. *Phys. Rev. B* **2007,** *75* (15), 153401.

Graphene composites with synthetic polymers

CHAPTER OUTLINE

7.1 Summary 141
7.2 Introduction 142
7.3 In situ Synthesis of Graphene/Polymer Composites 144
    7.3.1 Design and Preparation 144
    7.3.2 Characterization 144
    7.3.3 Properties and Applications 146
7.4 Post-modification Approaches 147
    7.4.1 Design and Preparation 147
    7.4.2 Characterization 147
    7.4.3 Properties and Applications 149
7.5 Conclusions 150
References 150

7.1 SUMMARY

Hybrid materials comprised of polymers and graphene are discussed in this chapter. The discovery that graphene/polymer composites could be made successfully has caused heightened interest in the multibillion-dollar polymer industry. The graphene in these composites provides additional mechanical strength and, more importantly, electronic properties such as conductivity allows the creation of new, functional polymer materials. The advanced features of synthetic polymers can be combined with the unique properties of graphene to produce composites whose properties can be tailored based on composition as well as the nature of synthetic polymer used. The one-dimensional characteristics of the polymer chains and the two-dimensional nature of graphene sheets are combined to produce mixed composites which provide unique opportunities for a variety of applications. The composites, therefore, are expected to have superior properties when compared to either of the pure components.

Here, we focus on the composites made from pristine graphene rather than graphene oxide (GO)- or reduced graphene oxide (rGO)-based hybrids. Stabilizing interactions at the polymer−graphene interface involve hydrophobic forces and predominantly, $\pi-\pi$ stacking. Our discussion here is divided into in situ and postmodification preparation methods. Polymers are more amenable for the direct preparation of polymer/graphene composites from pure graphite when the polymer component has a surface energy with reasonable match with that of graphene. The three-dimensional nature of the polymer-supported graphene in the matrix prevents restacking of the graphene layers. Different postmodifications can be executed by carefully designing or choosing the polymers that will stabilize the graphene—polymer interface. The major advantage of postmodification approaches is that the commercially available graphene and polymers can be further tailored as needed for specific applications. Purchasing the reagents saves time of synthesis, purification, and characterization of the individual components prior to the composite synthesis. These hybrids, which integrate the viscoelastic nature of the polymer and conductivity/strength of the graphene, are generally designed for applications in flexible and thin electronic devices. We conclude this chapter with a discussion on mass production opportunities of such composites in the future, as well as the (bio)degradation of the composite materials, an important criterion for environmentally sustainable growth of the field.

7.2 INTRODUCTION

Polymer composites of graphene and its derivatives have been given a large amount of attention since the discovery of graphene.[1] The structure—property relationships of polymers is well established and often utilized to generate polymeric materials with desirable properties such as hardness, flexibility, transparency, etc.[2] Polymers are part of our everyday life, from food packaging to organ transplants. Graphene/polymer hybrids have unique mechanical and optoelectronic properties, resulting in multifunctional materials that are attractive for fundamental research as well as for industry.[3] Graphene-reinforced polymer composites have made use of the mechanical strength of graphene, and found this strength to be tunable by varying the fraction of graphene in the composite.[4] Major interest in this field is driven by the electronic properties of graphene.[5] Composites with conductivity due to the graphene phase and flexibility from the polymer phase are of interest for applications in the electronics industry due to the possibility of flexible and printable circuits.[6] These composites are stabilized by covalent or noncovalent interactions at the polymer—graphene interfaces.[1]

Graphene can be incorporated into a polymer matrix during the polymerization step (in situ) or by conjugation after polymerization, which is known as postsynthetic synthesis (Fig. 7.1). The in situ polymer synthesis on the graphene surface could ensure maximum interfacial interaction between the two components, where the polymer growth on the two-dimensional sheets is

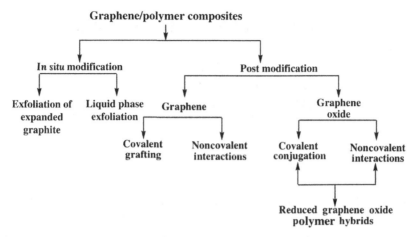

FIGURE 7.1

Major modification routes for graphene/polymer hybrid materials. In this chapter, we are not discussing the GO-based hybrids (right-most stem of the chart).

observed.[7] Postsynthetic methods are good alternatives if graphene interferes with reactions during polymerization, such as radical trapping in a free radical-mediated chain reaction.[8] Composites of commercially available polymers and graphene can be achieved by postsynthetic methods, which may be suitable for mass production. Postmodifications can be executed in a variety of solvent systems including water, whereas in situ polymerization conditions are generally limited to high-purity organic solvents. Polyethylene glycol (PEG)-modified graphene derivatives are synthesized in water, making them useful for biological applications, including catalysis and drug delivery.[9] Covalently modified graphene—polymer composites are prepared by harnessing the carbonyl chemistry of functional groups on the GO or the rGO surface.[10]

GO is popular among polymer chemists for composite preparation because of its solution processability and rich functional group chemistry.[11] Electrostatic interactions between negatively charged GO and positively charged polymers are effectively used for assemblies of GO and cationic polymers.[12] Subsequent reduction of GO to rGO within the polymer matrix improves the electronic conductivity of the composite.[11] The synthesis and applications of composites of GO and polymers have been extensively reviewed previously.[13,14] Herein, we continue to focus on pristine graphene-based composites, however relevant biological applications of GO-based hybrids are included. Pristine graphene can be tethered to polymers by a polymerization reaction on the graphene surface or by exfoliation of graphite using polymers as exfoliating agents (Fig. 7.1). These in situ preparations are discussed first, followed by postsynthetic approaches.

7.3 IN SITU SYNTHESIS OF GRAPHENE/POLYMER COMPOSITES

Polymer-enhanced graphite exfoliation is a convenient method for in situ syntheses of these hybrids. The key to successfully forming a stable graphene/polymer matrix is balancing the attractive and repulsive interactions of the polymer backbone or side group with that of the interactions compatible with graphene. Such approaches are discussed below.

7.3.1 DESIGN AND PREPARATION

Direct exfoliation of graphite or expanded graphite using solutions of polymers is the major route for the in situ preparation of graphene/polymer hybrids (Table 7.1). Exfoliation of graphite is achieved by ultrasonication or high-speed shearing where the polymer macromolecules prevent reaggregation of the graphene layers (Fig. 7.2).[15,16] Efficient exfoliations are attained when the solubility parameter of the graphene, polymer, and the solvent matches.[17] The interface of the polymer/graphene is stabilized by hydrophobic and/or $\pi-\pi$ interactions, where polymers with aromatic groups, such as polystyrene, gave composites of high stability.[18] Moreover, pyrene-modified monomers are designed for effective $\pi-\pi$ stacking of the polymer chains, when aromatic moieties are not available in the polymer backbone.[19] Amphiphilic copolymers with periodic aromatic and nonaromatic groups are interesting candidates for graphene/polymer hybrids that would be stable in water.[16] One disadvantage of this approach is that the solvent for polymerization must also be suitable for exfoliation, restricting it to a narrow choice of solvents. Direct exfoliation is also studied with amphiphilic protein molecules, which were discussed in Chapter 2 (Synthetic routes to graphene preparation from the perspectives of possible biological applications), showing that they are suitable for making graphene/biopolymer composites. The synthetic polymer-based graphene composites and their properties are listed in Table 7.1

7.3.2 CHARACTERIZATION

Graphene/polymer composites are characterized by standard techniques described previously in Chapter 3: (Characterization techniques for graphene). Additionally, physical characterization of the polymer component is also carried out. Most of the above-discussed studies used graphene as an additive to the polymer matrix for enhancement in conductivity and mechanical strength.[20] Chemical composition, morphology, distribution of the two phases, characterization of the phase boundaries, defect sites, and their abundance are some of the important characteristics to examine. Thermogravimetric analysis (TGA), stress—strain measurements, and rheological studies are essential to compare the properties of the polymer with and without the graphene additive. Atomic force microscopy

Table 7.1 Synthesis and Properties of In Situ Functionalized Graphene/ Polymer Hybrids

System	Type of Graphene	Properties and Applications
Pyrene functionalized block copolymer/ graphene[20]	Direct exfoliation in the polymer	Tunable electrical conductivity and tensile strength
Polystyrene/graphene[21]	Direct exfoliation in the polymer	Conductive foams
Block copolymer/ graphene[22]	Direct exfoliation in the polymer	145% increase in storage modulus by 0.1% weight fraction of graphene
Block copolymer/ graphene[23]	Direct exfoliation in the polymer	Photo-induced electron transfer from porphyrin to graphene
Block copolymer/ graphene[24]	Direct exfoliation in the polymer	High aqueous stability
Block copolymer/ graphene[25]	Direct exfoliation in the polymer	Biocompatible pluronic and tetronic block copolymers were used
Vinylimidazole-based polymer/graphene[26]	Direct exfoliation in the polymer	High electrical conductivity (3.6×10^3 S/m) and cytotoxicity (90% cell viability)
Poly(3-hexylthiophene)/ graphene[27]	Direct exfoliation in the polymer	Conductive ink
Ethyl cellulose/ graphene[28]	Direct exfoliation in the polymer	Conductive ink
Poly(vinylpyrrolidone)/ graphene[29]	Direct exfoliation in the polymer	Methanol oxidation performance better than rGO-based composite
Quinquethiophene-terminated PEG/ graphene[30]	Direct exfoliation in the polymer	Conductive-transparent thin film
Oligomeric ionic liquid/ graphene[31]	Direct exfoliation by microwave irradiation	95% single-layer graphene
Pyrene-DNA/graphene	Direct exfoliation in DNA	Stable aqueous solution of graphene[32]
Epoxy resin/graphite[33]	Expanded graphite or GO	Flame retardant material: $\sim 40\%$ reduction in flammability
Polyethylene/graphite[34]	Expanded graphite (H_2SO_4/HNO_3)	Expansion of interlayer distance from 0.336 to 28.15 nm (XRD)
Polystyrene/graphite[35]	Expanded graphite (H_2SO_4/HNO_3)	Conductive polystyrene nanocomposite

(AFM), transmission electron microscopy (TEM), and scanning electron microscopy (SEM) techniques are also used to elucidate the morphology of the matrix, especially when the block copolymers form self-assembled brush-like structures.[25] Because there are well-established techniques to characterize polymers and graphene, it is generally not difficult to evaluate these hybrid materials but

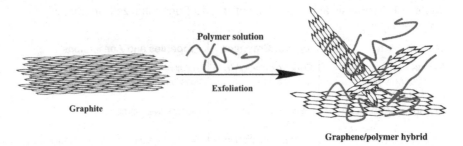

FIGURE 7.2

Direct exfoliation of graphite with polymers for one-pot preparation of graphene/polymer hybrids.

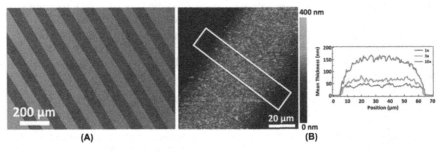

FIGURE 7.3

(A) Ink jet printed pattern of graphene exfoliated in ethyl cellulose. (B) AFM image of the single strip and its height profile after subsequent printing ($1\times$, $3\times$, and $10\times$). The resistance of the circuit was as low as 4 mΩ.cm (© American Chemical Society).[37]

characterization of the phase boundaries or grain boundaries and defect site mapping can be challenging. An ideal composite will have a uniform composition with uniform characteristics.

7.3.3 PROPERTIES AND APPLICATIONS

Dramatic unification of the properties of graphene and polymers has been observed in these hybrids. Graphene imparts mechanical stability/strength and electronic conductivity to the hybrid, while polymers stabilize graphene sheets from reaggregation. Commercially available graphene dispersions are usually stabilized using polymers (Sigma-Aldrich, Graphenea.com, etc.). Printable, conductive inks are one of the major focuses of this research, where the viscoelastic nature of the polymer solution is combined with electronic conductivity of exfoliated graphene (Fig. 7.3).[6] A polymer matrix also ensures the electrical contact between the individual graphene flakes, which minimizes ohmic resistance of the hybrid. Use of conductive polymers, such as thiophene polymers, further enhances the electronic contact throughout the matrix.[27] Recently, these polymers have also been utilized as 3-D printing materials where conductivity can be an added feature to the objects of interest, creating a new arena for next-generation electronics.[36]

7.4 POST-MODIFICATION APPROACHES

Polymers can be tethered to graphene layers by chemical conjugation or by non-covalent interactions. The $\pi-\pi$ stacking ability of the graphene surface is effectively used here.[38] Covalent chemistry of carbonyl groups is used in the case of GO-based hybrids. We restrict our focus to pristine graphene and, therefore, most of the GO-based hybrids are not included here. The approaches are discussed below.

7.4.1 DESIGN AND PREPARATION

Assembly of polymers on the graphene surface can be achieved through surface chemistry and solution-phase approaches. Graphene synthesized by CVD, or by direct exfoliation of graphite, is used for the construction of these assemblies (Table 7.2). The advantage of this approach is that graphene can be exfoliated under favorable conditions, which are well established and thoroughly characterized (discussed in Chapter 2: Synthetic routes to graphene preparation from the perspectives of possible biological applications).[39] Likewise, the polymers can also be synthesized and purified individually prior to assembly. The commercial availability of most of these polymers, such as pyrene-modified PEG, favors the mass production of the hybrids (Fig. 7.4).[40] PEG-modified graphene derivatives are widely used in biological studies for drug delivery and biosensing because of their stability in biological media, due to the PEG, and the high surface area platform they afford.[9] The methods and applications of polymer-modified graphene hybrids are listed below (Table 7.2).

7.4.2 CHARACTERIZATION

Postmodification of graphene with the synthetic polymer is done after the purification and characterization of individual components. Thus, analysis of these hybrids is focused on finding the final morphology of the material after the modification. TEM, SEM, and AFM studies are usually performed to ensure the polymer conjugation and to visualize the separate graphene sheets in the

Table 7.2 Synthesis and Properties of Postmodified Graphene/Polymer Hybrids

System	Type of Graphene/ Method	Properties and Applications
Ethyl cellulose/graphene[41]	Graphite exfoliated in DMF and redispersed in polymer	Inject printable conductive ink
Polycarbonate/graphene[42]	Graphite exfoliated in NMP and redispersed in polymer	3-D printing material
Epoxy/graphene[43]	CVD-grown single-layer graphene	Scratch-resistant material
PEG/graphene[44]	• CVD-grown single-layer graphene • Pyrene-modified PEG for assembly	FET-based cancer biomarker detection
• Poly(methyl methacrylate)/ graphene • Poly(glycidyl methacrylate)/ graphene • Poly(N-isopropylacrylamide)/ graphene	• CVD-grown graphene • Initiator was attached to graphene using a pyrene linker • Polymerization was done by immersing in catalyst solution	DNA biosensors
Poly[(2-(methacryloyloxy)ethyl) trimethylammonium chloride] (PMETAC)/graphene[45]		
Polystyrene/graphene[46]	• CVD-grown graphene • Photo-initiated free radical polymerization of styrene on graphene surface	Brush polymers on graphene
Block copolymer/Graphene[47]	CVD-grown graphene	Solvent switchable field effect transistors (FETs)
β-Cyclodextrin/graphene[48]	Graphene by pyrolysis	High aqueous stability of graphene
Crosslinked silicon polymer/ graphene[49]	Graphite exfoliated in NMP	Electromechanical motion sensor
DNA/GO[50]	GO by Hummer's method	Self-healing hydrogels
Poly(acrylic acid)/GO[51]	GO by Hummer's method	Protective shield for enzyme catalysis at denaturing conditions
PEG/GO[52]	• GO by Hummer's method • Covalent attachment of PEG by amide coupling	Delivery for anticancer drug in cells

matrix.[45] Phase separation of graphene and the polymer is possible by uncontrolled aggregation or restacking, which can be ruled out after microscopy analysis. TGA is another commonly used technique to monitor the change in glass transition temperature of the polymer after conjugation with graphene.[43]

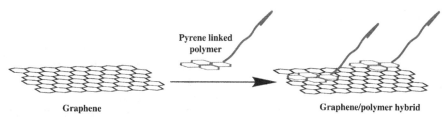

FIGURE 7.4

Pyrene-modified polymers for making graphene/polymer hybrids stabilized by $\pi-\pi$ stacking interaction between graphene and pyrene.

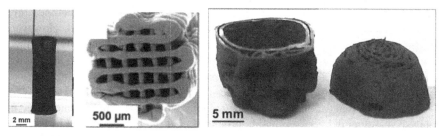

FIGURE 7.5

3-D printed composite of polylactide-*co*-glycolide (85:15) copolymer and exfoliated graphene (3–8 layers), which showed in vivo biocompatibility (© American Chemical Society).[53]

Conductivity measurements, stress—strain analysis, opacity, etc. are carried out when the composite is targeted for applications in stretchable transparent electronic circuits.

7.4.3 PROPERTIES AND APPLICATIONS

Conjugated polymer-graphene hybrids have a wide range of applications, from electronics to biomedicine (Table 7.2). The relatively easy synthetic modifications and purified precursors make this approach simpler than the in situ methods. Moreover, aqueous processability in some methods is beneficial for biological studies. Electronic applications of these hybrids use the viscosity and thermosetting property of the polymer for printing circuits by either using an ink jet printer or a 3-D printer (Fig. 7.5).[53] Water-dispersible, PEG-modified graphene was used for delivery of the hydrophobic anticancer drug (paclitaxel), which hydrophobically binds to the graphene surface.[52] We have demonstrated that polyacrylic acid and GO synergistically provide stability to enzymes for biocatalysis under biologically unfavorable conditions of temperature, pH, and surfactant concentrations.[51] Such applications are a promising route for real-life uses of graphene.

7.5 CONCLUSIONS

The graphene/polymer composites combine the electronic and mechanical properties of graphene with the properties of the synthetic polymers.[41] Conversely, graphene can also promote the mechanical strength of polymer composites in the binary system. For example, the 2-D component could provide an effective barrier against gas permeation or for filtration applications. The combination of the 1-D and 2-D molecular components in the hybrid is an interesting feature, which is further augmented with the contrasting physical/chemical properties of the two components. Exfoliated graphite in polymer solutions forms stable colloids, attributed to the 3-D structure of the adsorbed polymer molecules.[17] The in situ exfoliation methods have the possibility of inexpensive mass production, where graphite, the polymer, and the solvent, are the precursors. Current methods rely on ultrasonication to delaminate graphite, which results in poor yield. However, implementation of shear or ball milling-based exfoliation methods shows much promise to overcome this limitation.[15] Biomolecules, such as proteins, are known to stabilize graphene in solution, which may allow for a more sustainable route for graphene suspensions of biocomposites to target the biological markets.[54]

Biomedical application of graphene/polymer composites and applications in catalysis and drug delivery are highly promising, but the biodegradation of the graphene and its environmental fate are not well investigated.[55,56] Partial degradation of graphene sheets by enzymatic activities could lead to the release of polyatomic hydrocarbons, which are known carcinogens and mutagens.[56] Controlling these undesired outcomes using polymer conjugation is a challenge to be addressed. Also, GO has oxidative debris and basal plane defects that are prone to degradation by enzymes. The use of exfoliated or CVD-grown graphene may, therefore, be beneficial for storage/usage but environmental fate is an important consideration.[57,58] Degradation of the polymer can also release toxic substances, or cause the sonolysis of PEG, resulting in the release of ethylene glycol during composite preparation.[59] Thus, biomedical applications of these materials might require a critical design strategy that balances the benefits as well as remediation or biological clearance. Polymer hybrids of inorganic layered materials such as hexagonal boron nitride,[60] MoS_2,[61] WS_2,[62] and black phosphorus[63] are also known, not discussed here. Hybrids of two-dimensional materials show great promise where the structure - property relationship between the 2-D material and the polymer can be tuned, allowing for the creation of tailored functional materials.

REFERENCES

1. Kuilla, T.; Bhadra, S.; Yao, D.; Kim, N. H.; Bose, S.; Lee, J. H. Recent Advances in Graphene Based Polymer Composites. *Prog. Polymer Sci.* **2010,** *35* (11), 1350−1375.
2. Andrews, E. H. Structure-Property Relationships in a Polymer. *Angew. Chem. Int. Ed. Engl.* **1974,** *13* (2), 113−121.

3. Loh, K. P.; Bao, Q.; Ang, P. K.; Yang, J. The Chemistry of Graphene. *J. Mater. Chem.* **2010,** *20* (12), 2277−2289.

4. Terrones, M.; Martín, O.; González, M.; Pozuelo, J.; Serrano, B.; Cabanelas, J. C.; Vega-Díaz, S. M.; Baselga, J. Interphases in Graphene Polymer-based Nanocomposites: Achievements and Challenges. *Adv. Mater.* **2011,** *23* (44), 5302−5310.

5. Guo, S.; Dong, S. Graphene Nanosheet: Synthesis, Molecular Engineering, Thin Film, Hybrids, and Energy and Analytical Applications. *Chem. Soc. Rev.* **2011,** *40* (5), 2644−2672.

6. Torrisi, F.; Hasan, T.; Wu, W.; Sun, Z.; Lombardo, A.; Kulmala, T. S.; Hsieh, G.-W.; Jung, S.; Bonaccorso, F.; Paul, P. J. Inkjet-Printed Graphene Electronics. *ACS Nano* **2012,** *6* (4), 2992−3006.

7. Das, S.; Wajid, A. S.; Shelburne, J. L.; Liao, Y.-C.; Green, M. J. Localized *In Situ* Polymerization on Graphene Surfaces for Stabilized Graphene Dispersions. *ACS Appl. Mater. Interfaces* **2011,** *3* (6), 1844−1851.

8. Kolanthai, E.; Bose, S.; Bhagyashree, K.; Bhat, S.; Asokan, K.; Kanjilal, D.; Chatterjee, K. Graphene Scavenges Free Radicals to Synergistically Enhance Structural Properties in a Gamma-Irradiated Polyethylene Composite Through Enhanced Interfacial Interactions. *Phys. Chem. Chem. Phys.* **2015,** *17* (35), 22900−22910.

9. Wang, Y.; Li, Z.; Wang, J.; Li, J.; Lin, Y. Graphene and Graphene Oxide: Biofunctionalization and Applications in Biotechnology. *Trends Biotechnol.* **2011,** *29* (5), 205−212.

10. Chen, D.; Feng, H.; Li, J. Graphene Oxide: Preparation, Functionalization, and Electrochemical Applications. *Chem. Rev.* **2012,** *112* (11), 6027−6053.

11. Dreyer, D. R.; Park, S.; Bielawski, C. W.; Ruoff, R. S. The Chemistry of Graphene Oxide. *Chem. Soc. Rev.* **2010,** *39* (1), 228−240.

12. Zou, J.; Kim, F. Diffusion Driven Layer-By-Layer Assembly of Graphene Oxide Nanosheets Into Porous Three-Dimensional Macrostructures. *Nat. Commun.* **2014,** *5,* 5254 (article number).

13. Potts, J. R.; Dreyer, D. R.; Bielawski, C. W.; Ruoff, R. S. Graphene-Based Polymer Nanocomposites. *Polymer. (Guildf).* **2011,** *52* (1), 5−25.

14. Salavagione, H. J.; Diez-Pascual, A. M.; Lazaro, E.; Vera, S.; Gomez-Fatou, M. A. Chemical Sensors Based on Polymer Composites with Carbon Nanotubes and Graphene: The Role of the Polymer. *J. Mater. Chem. A* **2014,** *2* (35), 14289−14328.

15. Hernandez, Y.; Nicolosi, V.; Lotya, M.; Blighe, F. M.; Sun, Z.; De, S.; McGovern, I. T.; Holland, B.; Byrne, M.; Gun'Ko, Y. K., et al. High-Yield Production of Graphene by Liquid-Phase Exfoliation of Graphite. *Nat. Nano* **2008,** *3* (9), 563−568.

16. Ciesielski, A.; Samorì, P. Graphene via Sonication Assisted Liquid-Phase Exfoliation. *Chem. Soc. Rev.* **2014,** *43* (1), 381−398.

17. May, P.; Khan, U.; Hughes, J. M.; Coleman, J. N. Role of Solubility Parameters in Understanding the Steric Stabilization of Exfoliated Two-Dimensional Nanosheets by Adsorbed Polymers. *J. Phys. Chem. C* **2012,** *116* (20), 11393−11400.

18. Lechner, C.; Sax, A. F. Adhesive Forces Between Aromatic Molecules and Graphene. *J. Phys. Chem. C* **2014,** *118* (36), 20970−20981.

19. Gatti, T.; Vicentini, N.; Mba, M.; Menna, E. Organic Functionalized Carbon Nanostructures for Functional Polymer-Based Nanocomposites. *Eur. J. Org. Chem.* **2016,** *2016* (6), 1071−1090.

20. Liu, Z.; Liu, J.; Cui, L.; Wang, R.; Luo, X.; Barrow, C. J.; Yang, W. Preparation of Graphene/Polymer Composites by Direct Exfoliation of Graphite in Functionalised Block Copolymer Matrix. *Carbon. N. Y.* **2013,** *51,* 148−155.

21. Woltornist, S. J.; Carrillo, J.-M. Y.; Xu, T. O.; Dobrynin, A. V.; Adamson, D. H. Polymer/Pristine Graphene Based Composites: From Emulsions to Strong, Electrically Conducting Foams. *Macromolecules.* **2015,** *48* (3), 687−693.

22. Popescu, M.-T.; Tasis, D.; Papadimitriou, K. D.; Gkermpoura, S.; Galiotis, C.; Tsitsilianis, C. Colloidal Stabilization of Graphene Sheets by Ionizable Amphiphilic Block Copolymers in Various Media. *RSC Adv.* **2015,** *5* (109), 89447−89460.

23. Skaltsas, T.; Pispas, S.; Tagmatarchis, N. Photoinduced Charge-Transfer Interactions on a Graphene/Block Copolymer Electrostatically Bound to Tetracationic Porphyrin in Aqueous Media. *Chem.−A Eur. J.* **2013,** *19* (28), 9286−9290.

24. Skaltsas, T.; Karousis, N.; Yan, H.-J.; Wang, C.-R.; Pispas, S.; Tagmatarchis, N. Graphene Exfoliation in Organic Solvents and Switching Solubility in Aqueous Media with the Aid of Amphiphilic Block Copolymers. *J. Mater. Chem.* **2012,** *22* (40), 21507−21512.

25. Seo, J.-W. T.; Green, A. A.; Antaris, A. L.; Hersam, M. C. High-Concentration Aqueous Dispersions of Graphene Using Nonionic, Biocompatible Block Copolymers. *J. Phys. Chem. Lett.* **2011,** *2* (9), 1004−1008.

26. Cui, J.; Song, Z.; Xin, L.; Zhao, S.; Yan, Y.; Liu, G. Exfoliation of Graphite to Few-Layer Graphene in Aqueous Media with Vinylimidazole-Based Polymer as High-Performance Stabilizer. *Carbon. N. Y.* **2016,** *99,* 249−260.

27. Iguchi, H.; Higashi, C.; Funasaki, Y.; Fujita, K.; Mori, A.; Nakasuga, A.; Maruyama, T. Rational and Practical Exfoliation of Graphite Using Well-Defined Poly (3-Hexylthiophene) for the Preparation of Conductive Polymer/Graphene Composite. *Sci. Rep.* **2017,** *7,* 39937.

28. Liang, Y. T.; Hersam, M. C. Highly Concentrated Graphene Solutions via Polymer Enhanced Solvent Exfoliation and Iterative Solvent Exchange. *J. Am. Chem. Soc.* **2010,** *132* (50), 17661−17663.

29. Wang, H.; Xia, B.; Yan, Y.; Li, N.; Wang, J.-Y.; Wang, X. Water-Soluble Polymer Exfoliated Graphene: As Catalyst Support and Sensor. *J. Phys. Chem. B.* **2013,** *117* (18), 5606−5613.

30. Kang, M. S.; Kim, K. T.; Lee, J. U.; Jo, W. H. Direct Exfoliation of Graphite Using a Non-Ionic Polymer Surfactant for Fabrication of Transparent and Conductive Graphene Films. *J. Mater. Chem. C* **2013,** *1* (9), 1870−1875.

31. Matsumoto, M.; Saito, Y.; Park, C.; Fukushima, T.; Aida, T. Ultrahigh-Throughput Exfoliation of Graphite into Pristine 'Single-Layer'graphene Using Microwaves and Molecularly Engineered Ionic Liquids. *Nat. Chem.* **2015,** *7* (9), 730−736.

32. Liu, F.; Choi, J. Y.; Seo, T. S. DNA Mediated Water-Dispersible Graphene Fabrication and Gold Nanoparticle-Graphene Hybrid. *Chem. Commun.* **2010,** *46* (16), 2844−2846.

33. Guo, Y.; Bao, C.; Song, L.; Yuan, B.; Hu, Y. *In Situ* Polymerization of Graphene, Graphite Oxide, and Functionalized Graphite Oxide into Epoxy Resin and Comparison Study of on-the-Flame Behavior. *Ind. Eng. Chem. Res.* **2011,** *50* (13), 7772−7783.

34. Fim, Fd. C.; Guterres, J. M.; Basso, N. R.; Galland, G. B. Polyethylene/Graphite Nanocomposites Obtained by *In Situ* Polymerization. *J. Polymer Sci. Part A Polymer Chem.* **2010,** *48* (3), 692−698.

35. Chen, G.; Wu, C.; Weng, W.; Wu, D.; Yan, W. Preparation of Polystyrene/Graphite Nanosheet Composite. *Polymer. (Guildf).* **2003,** *44* (6), 1781−1784.
36. Wei, X.; Li, D.; Jiang, W.; Gu, Z.; Wang, X.; Zhang, Z.; Sun, Z. 3D Printable Graphene Composite. *Sci. Rep.* **2015,** *5*, 11181.
37. Secor, E. B.; Prabhumirashi, P. L.; Puntambekar, K.; Geier, M. L.; Hersam, M. C. Inkjet Printing of High Conductivity, Flexible Graphene Patterns. *J. Phys. Chem. Lett.* **2013,** *4* (8), 1347−1351.
38. Björk, J.; Hanke, F.; Palma, C.-A.; Samori, P.; Cecchini, M.; Persson, M. Adsorption of Aromatic and Anti-Aromatic Systems on Graphene through $\pi - \pi$ Stacking. *J. Phys. Chem. Lett.* **2010,** *1* (23), 3407−3412.
39. Nicolosi, V.; Chhowalla, M.; Kanatzidis, M. G.; Strano, M. S.; Coleman, J. N. Liquid Exfoliation of Layered Materials. *Science* **2013,** *340* (6139), 1226419.
40. Björk, J.; Hanke, F.; Palma, C.-A.; Samori, P.; Cecchini, M.; Persson, M. Adsorption of Aromatic and Anti-Aromatic Systems on Graphene Through $\pi - \pi$ Stacking. *J. Phys. Chem. Lett.* **2010,** *1* (23), 3407−3412.
41. Li, J.; Ye, F.; Vaziri, S.; Muhammed, M.; Lemme, M. C.; Östling, M. Efficient Inkjet Printing of Graphene. *Adv. Mater.* **2013,** *25* (29), 3985−3992.
42. Lago, E.; Toth, P. S.; Pugliese, G.; Pellegrini, V.; Bonaccorso, F. Solution Blending Preparation of Polycarbonate/Graphene Composite: Boosting the Mechanical and Electrical Properties. *RSC Adv.* **2016,** *6* (100), 97931−97940.
43. Xiao, X.; Xie, T.; Cheng, Y.-T. Self-Healable Graphene Polymer Composites. *J. Mater. Chem.* **2010,** *20* (17), 3508−3514.
44. Gao, N.; Gao, T.; Yang, X.; Dai, X.; Zhou, W.; Zhang, A.; Lieber, C. M. Specific Detection of Biomolecules in Physiological Solutions Using Graphene Transistor Biosensors. *Proc. Natl. Acad. Sci.* **2016,** *113* (51), 14633−14638.
45. Gao, T.; Ng, S.-W.; Liu, X.; Niu, L.; Xie, Z.; Guo, R.; Chen, C.; Zhou, X.; Ma, J.; Jin, W. Transferable, Transparent and Functional Polymer@ Graphene 2D Objects. *NPG Asia Mater.* **2014,** *6* (9), e130.
46. Steenackers, M.; Gigler, A. M.; Zhang, N.; Deubel, F.; Seifert, M.; Hess, L. H.; Lim, C. H. Y. X.; Loh, K. P.; Garrido, J. A.; Jordan, R. Polymer Brushes on Graphene. *J. Am. Chem. Soc.* **2011,** *133* (27), 10490−10498.
47. Liu, S.; Jamali, S.; Liu, Q.; Maia, J.; Baek, J.-B.; Jiang, N.; Xu, M.; Dai, L. Conformational Transitions of Polymer Brushes for Reversibly Switching Graphene Transistors. *Macromolecules.* **2016,** *49* (19), 7434−7441.
48. Zhou, W.; Li, W.; Xie, Y.; Wang, L.; Pan, K.; Tian, G.; Li, M.; Wang, G.; Qu, Y.; Fu, H. Fabrication of Noncovalently Functionalized Brick-Like [Small Beta]-Cyclodextrins/Graphene Composite Dispersions with Favorable Stability. *RSC Adv.* **2014,** *4* (6), 2813−2819.
49. Boland, C. S.; Khan, U.; Ryan, G.; Barwich, S.; Charifou, R.; Harvey, A.; Backes, C.; Li, Z.; Ferreira, M. S.; Möbius, M. E., et al. Sensitive Electromechanical Sensors Using Viscoelastic Graphene-Polymer Nanocomposites. *Science* **2016,** *354* (6317), 1257−1260.
50. Xu, Y.; Wu, Q.; Sun, Y.; Bai, H.; Shi, G. Three-Dimensional Self-Assembly of Graphene Oxide and DNA into Multifunctional Hydrogels. *ACS Nano* **2010,** *4* (12), 7358−7362.
51. Zore, O. V.; Pattammattel, A.; Gnanaguru, S.; Kumar, C. V.; Kasi, R. M. Bienzyme−Polymer−Graphene Oxide Quaternary Hybrid Biocatalysts: Efficient

Substrate Channeling Under Chemically and Thermally Denaturing Conditions. *ACS Catalysis* **2015,** *5* (9), 4979−4988.

52. Liu, Z.; Robinson, J. T.; Sun, X.; Dai, H. PEGylated Nano-Graphene Oxide for Delivery of Water Insoluble Cancer Drugs. *J. Am. Chem. Soc.* **2008,** *130* (33), 10876.

53. Jakus, A. E.; Secor, E. B.; Rutz, A. L.; Jordan, S. W.; Hersam, M. C.; Shah, R. N. Three-Dimensional Printing of High-Content Graphene Scaffolds for Electronic and Biomedical Applications. *ACS Nano* **2015,** *9* (4), 4636−4648.

54. Pattammattel, A.; Kumar, C. V. Kitchen Chemistry 101: Multigram Production of High Quality Biographene in a Blender with Edible Proteins. *Adv. Funct. Mater.* **2015,** *25* (45), 7088−7098.

55. Song, Z.; Xu, Y.; Yang, W.; Cui, L.; Zhang, J.; Liu, J. Graphene/Tri-Block Copolymer Composites Prepared via RAFT Polymerizations for Dual Controlled Drug Delivery via pH STIMULATION and biodegradation. *Eur. Polym J.* **2015,** *69*, 559−572.

56. Bai, H.; Jiang, W.; Kotchey, G. P.; Saidi, W. A.; Bythell, B. J.; Jarvis, J. M.; Marshall, A. G.; Robinson, R. A.; Star, A. Insight into the Mechanism of Graphene Oxide Degradation via the Photo-Fenton Reaction. *J. Phys. Chem. C. Nanomater. Interfaces* **2014,** *118* (19), 10519.

57. Murray, E.; Thompson, B. C.; Sayyar, S.; Wallace, G. G. Enzymatic Degradation of Graphene/Polycaprolactone Materials for Tissue Engineering. *Polymer Degrad. Stab.* **2015,** *111*, 71−77.

58. Pattammattel, A.; Williams, C. L.; Pande, P.; Tsui, W. G.; Basu, A. K.; Kumar, C. V. Biological Relevance of Oxidative Debris Present in as-Prepared Graphene Oxide. *RSC Adv.* **2015,** *5* (73), 59364−59372.

59. Murali, V. S.; Wang, R.; Mikoryak, C. A.; Pantano, P.; Draper, R. Rapid Detection of Polyethylene Glycol Sonolysis Upon Functionalization of Carbon Nanomaterials. *Exp. Biol. Med.* **2015,** *240* (9), 1147−1151.

60. Li, T.-L.; Hsu, S. L.-C. Enhanced Thermal Conductivity of Polyimide Films via a Hybrid of Micro-and Nano-Sized Boron Nitride. *J. Phys. Chem. B.* **2010,** *114* (20), 6825−6829.

61. Yang, L.; Wang, S.; Mao, J.; Deng, J.; Gao, Q.; Tang, Y.; Schmidt, O. G. Hierarchical MoS2/Polyaniline Nanowires with Excellent Electrochemical Performance for Lithium-Ion Batteries. *Adv. Mater.* **2013,** *25* (8), 1180−1184.

62. Vega-Mayoral, V.; Backes, C.; Hanlon, D.; Khan, U.; Gholamvand, Z.; O'Brien, M.; Duesberg, G. S.; Gadermaier, C.; Coleman, J. N. Photoluminescence from Liquid-Exfoliated WS2 Monomers in Poly (Vinyl Alcohol) Polymer Composites. *Adv. Funct. Mater.* **2016,** *26* (7), 1028−1039.

63. Li, D.; Castillo, A. E. D. R.; Jussila, H.; Ye, G.; Ren, Z.; Bai, J.; Chen, X.; Lipsanen, H.; Sun, Z.; Bonaccorso, F. Black Phosphorus Polycarbonate Polymer Composite for Pulsed Fibre Lasers. *Appl. Mater. Today* **2016,** *4*, 17−23.

Graphene composites with proteins and biologics

CHAPTER OUTLINE

8.1 Summary 155
8.2 Introduction 156
8.3 Design and Preparation of Graphene Biohybrids 159
    8.3.1 Covalent and Noncovalent Approaches 159
8.4 Characterization of Composites..162
    8.4.1 Equilibrium Binding Studies...163
    8.4.2 Fluorescence Spectroscopy 165
    8.4.3 Zeta Potential Studies...166
    8.4.4 Atomic Force Microscopy 167
    8.4.5 X-Ray Diffraction 168
    8.4.6 CD Spectroscopy 168
    8.4.7 Biological Activity Studies 171
8.5 Biocatalysis Using Graphene Platform 171
    8.5.1 Preservation of Structure and Function of Bound Enzymes.................172
    8.5.2 Thermodynamic Stabilities of Enzymes...174
    8.5.3 Kinetic Stability of Enzymes 175
8.6 Biosensing...177
    8.6.1 Fluorescence-Based Graphene Biosensors 177
    8.6.2 Electrochemical Graphene Sensors 178
8.7 Drug Delivery Applications...179
8.8 Conclusions...179
References 180

8.1 SUMMARY

Biological applications of graphene derivatives are discussed in this chapter. The discussion focuses on graphene oxide (GO) hybrid systems because of their good solubility in water as well as their facile availability. General interactions at the bio—nano interface will be examined first, and facilitate the discussion of the more complex interface of biological systems at nanosurfaces. The experimental techniques used here are different from previously discussed methods of graphene characterization, and these methodologies will be discussed using the examples of biohybrids developed in our lab. We stress the need for minimizing adverse

Introduction to Graphene. DOI: http://dx.doi.org/10.1016/B978-0-12-813182-4.00008-8

interactions at the bio—nano interface if one were to preserve the biological function of the biomolecule of interest, which is in contact with the graphene substrate. Applications of the graphene biointerface in biocatalysis are discussed next. These include the "stable-on-the-table" enzyme hybrids, discovered in our laboratory, that are functional under challenge. Molecular engineering of this interface by systematic methods resulted in enzyme systems which are stable for over a month at elevated temperatures. Applications of these biohybrid materials for the detection of important molecules, such as glucose, below physiological levels are discussed, where graphene acts as a signal transducer and electrode. Highly promising drug-delivery systems based on GO are discussed at the end. A discussion about possible future efforts to create graphene with better biocompatibility and stability under physiological conditions, which will be imperative for application of these systems into clinical systems, will conclude this important chapter.

8.2 INTRODUCTION

Applications of nanomaterials in many exciting fields of research such as nanomedicine, nanotheraputics, and bionanosensors are being witnessed.[1] Since its invention, graphene with unique properties including high surface area ($1500-2000$ m^2/g), chemical/thermal stability, and many other optoelectronic properties, is suitable for the development of drug-delivery vehicles, biosensors, biocatalysis, and bioelectronics.[2-9] However, the poor dispersive nature of graphene in aqueous media obstructed the development of graphene for biological applications. The availability of protein-stabilized graphene suspensions in water, as reported in this book, is expected to change the status of this field in the coming years. Due to the scarcity of graphene/water dispersions, its surrogate, GO which is the oxidized form of graphene, is often used for biohybrid preparation and characterization. GO possesses oxygen functional groups on its surface and edges which provide surface charge and hydrophilic functional groups for stabilizing GO in water.[10] GO with hydroxyl, carboxyl, and carbonyl groups on its surface can be easily modified by chemical reactions to further tailor the properties of GO as desired.[11] It can be hard, however, to evaluate the interaction of GO with the bio—nano interface because of the random distribution of hydrophobic and hydrophilic groups on the flakes.

Molecular interactions between biomolecules and nanosurfaces are expected to be complex because of the presence of multiple functional groups, heterogeneity of the interface, and the interplay of multiple binding modes that contribute to the total interaction energy. Controlling these interactions is a significant challenge and some progress has been made toward this goal. Specific molecular moieties are attached to the graphene or GO surface to control the binding interactions as well as their location (Fig. 8.1). Streptavidin-avidin (or biotin),[12]

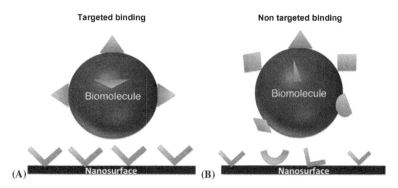

FIGURE 8.1

Complementary (A) and noncomplementary (B) binding of the biomacromolecule to the nanosheet to form the biohybrid. Complementary groups are introduced at the biomolecule and the nanosheet surfaces for targeted binding, whereas random binding of functional groups on the biomolecule to the nanosheet surface results in noncomplementary binding. Both methods produce biohybrids, but with contrasting requirements.

sugar-lectin,[13] antigen–antibody interactions, e.g., are thoroughly studied and commercially available for attaching biomolecules on to the GO surface. A major limitation of this approach of inducing targeted interactions between graphene and its biological partner, in addition to high cost, is the detailed chemical methodology that is required to attach the corresponding ligands and receptors.[14] Nontargeted and nonspecific interactions, on the other hand, do not require complementary ligands, and controlling these interactions is particularly challenging but they offer facile fabrication and mass production opportunities. The nonspecific interactions are often controlled by tuning the equilibrium conditions of charge, pH, and ionic strength.[15] Biomolecule–graphene interactions are designed by building electrostatic or weak noncovalent interactions between the partners of the biohybrid by appropriate modifications of the partners as needed.[16]

Binding approaches based on noncovalent interactions, such as electrostatic interactions, are convenient for the design, fabrication, and scale-up of nanobiohybrids. With this approach, no postmodifications are essential and both the nanosheet and the biomolecule possess electrostatic charges, hydrogen bonding donors and/or acceptors, hydrophobic sites, etc., which are suitable for a multivalent binding interaction (Fig. 8.2). The roles of solvent and counter ions at the charged bio–nano interfaces were investigated by our group using α-zirconium phosphate as a model nanosheet. This material has uniformly distributed surface phosphate groups, negatively charged at neutral pH, water-dispersible, and it has been proposed to bind proteins via an ion-coupled protein binding (ICPB) mechanism.[15] According to the ICPB mechanism, counter ions are released from the binding interfaces when a positively charged protein binds to a negatively charged

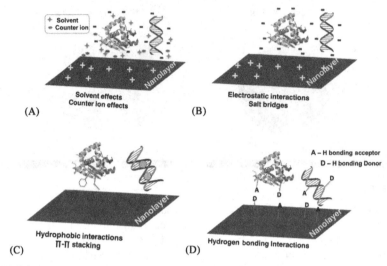

FIGURE 8.2

Nontargeted interactions at the bio—nanosheet interface.

nanosheet. On the other hand, the interaction of a negatively charged protein with a negatively charged nanosheet requires the sequestration of cations to the interface to reduce the pile-up of excess negative charge at the bio—nano interface. The immobilization of negatively charged DNA onto the anionic mica surface was only achieved in the presence of divalent metal ions, screening the unfavorable charge—charge repulsions, and these emphasize the importance of the ion sequestration mechanism at the bio—nano interface for maintaining a low-energy interface.[17] Strong attractive electrostatic interactions between the nanosheet and the biomolecule can also distort the secondary/tertiary structure of the biomolecule. Such distortion of structure could adversely influence the biological activity of the biomolecule.

The major goal when making the biohybrid is to conserve the function of the biomolecule under laboratory conditions, which are usually quite different from the native environment of the biomolecule in the physiological environment. Since optimum activity of biomolecules is required for applications in biomedical devices or for biocatalytic/sensing applications of the biohybrid, careful control of the bio—nano interactions deserves attention.[16] Carbon surfaces such as in carbon nanotubes, graphene, or GO are hydrophobic in nature, while most biomolecules are strongly hydrophilic.[18,19] Due to the hydrophobicity of the supporting nanosurface, proteins bound to the carbon surfaces tend to denature and unravel their secondary structure. This is because the interior functional groups of most proteins are generally hydrophobic and unraveling their structure will promote interactions with the hydrophobic carbon surface.[20] Unless well controlled, these interactions could deactivate the biomolecule altogether, unless this is desirable for a given application, this needs to be under chemical control. By controlled biofunctionalization of the carbon surface, protein denaturation at the interface can be minimized or avoided altogether.[21] Alternatively, biomolecules can be attached to the hydrophilic functional groups of GO by covalent linking, but there is still a certain extent of the exposure of hydrophobic regions of GO to the protein. The designs of the biohybrids via covalent and noncovalent surface functionalizations are described below.

8.3 DESIGN AND PREPARATION OF GRAPHENE BIOHYBRIDS

Graphene biohybrids are often designed to harness the biological function of the biomolecular systems while enhancing the mechanical and/or optoelectronic properties of graphene. Thus, the biohybrid is expected to have properties that are derived from both components, where each complements the other. However, this is more challenging than imagined because mutual interactions between the two components could alter the properties of one another. In this binary system, preserving the function of the biomolecule is of utmost importance, but preserving the optoelectronic properties of graphene could also be relevant, depending on the application at hand. Here, we describe the covalent and noncovalent methodologies that can be utilized to yield functional graphene biohybrids.

8.3.1 COVALENT AND NONCOVALENT APPROACHES

Covalent conjugation of biomolecules to GO is a straight-forward approach, where amide coupling chemistry is the most often used strategy.[22] Amide coupling chemistry, through the carbodiimide, is a favorable method and it efficiently couples the reactive amino groups on proteins with the carboxylic acid groups on GO to form the amide bonds.[23] Click chemistry of thiol-ene, azide-alkye, etc., is also explored after proper modification of the protein and graphene surfaces.[24] Biomacromolecules that are covalently attached to the carbon surfaces could also undergo denaturation due to a variety of noncovalent interactions between the biological molecule and the carbon surface. In the following sections, the effective use of the biointerface of graphene and its derivatives is described for the construction of biocatalysts, biosensors, and drug-delivery vehicles, using biomolecules as active materials. Table 8.1 summarizes the approaches for bioconjugation via covalent chemistry.

Noncovalent approaches are widely explored for biofunctionalizations at the bio−graphene interface.[34] Biomolecules, e.g., are assembled at the graphene surface via hydrogen bonding, electrostatic interactions, hydrophobic effect, and/or $\pi-\pi$ interactions. These same interactions are also employed to exfoliate graphite

Table 8.1 Covalent Approaches for Bioconjugation With Graphene Derivatives

System	Bioconjugation Method	Properties/Applications
CVD graphene–bovine serum albumin (BSA)[25]	Maleic anhydride modified to graphene by Diels–Alder reaction, followed by carbodiimide coupling of BSA	Quenching of fluorescence from Alexa dye modified BSA because of enhanced adsorption
GO–lipase[26]	Disulfide bond from thiosulfate on GO and cysteine in lipase	• About 90% retention in activity after conjugation • ~70% retention in activity after 10 recycling steps
GO–glucose oxidase (GOx)[27]	Nucleophilic addition of amines from protein to succinic anhydride-modified GO	Direct electron transfer from GOx to graphene-based electrode
GO–lipase[28]	Glutaraldehyde crosslinking	Biocatalysis at 70°C: 65% retention
GO–laccase[29]	Glutaraldehyde crosslinking	Approximately fivefold higher activity of GO hybrid at 60°C than free enzyme
GO–bovine serum albumin (BSA)[22]	Amide coupling of –COOH in GO and amine in the protein via carbodiimide chemistry	• Complete exfoliation of GO • Complete retention of electroactivity of BSA
GO–chitosan[30]	Amide coupling of –COOH in GO and amine in the protein via carbodiimide chemistry	The composite was used a protective layer against protease degradation of BSA
GO–*Cyanobacterium* metallothionein (SmtA)[31]	Amide coupling of –COOH in GO and amine in the protein via carbodiimide chemistry	Cysteine-rich metal-binding protein SmtA on GO removes trace cadmium from water
GO–DA-grafted heparin[32]	Amide coupling of –COOH in GO and amine in the protein via carbodiimide chemistry	High biocompatibility tested in human umbilical vein endothelial cells
GO–Angiopep-2[33]	GO-N_3 and alkyne-based peptide attached via Cu catalyzed click chemistry	• Angiopep-2 is a blood-brain barrier targeting protein • Low cell viability of the conjugates in human cancer cells (A549)

to graphene in water, as discussed in Chapter 3, Characterization techniques for graphene. This strategy does not require chemical crosslinkers or purification steps. Thus, exfoliation and binding are carried out in one step. Alternatively, the exfoliated graphene is then used to form biohybrids by contacting with solutions of biomolecule to fabricate biosensors and drug-delivery vehicles.[34] These methods are presented in Table 8.2.

Table 8.2 Approaches for Biofunctionalization of Graphene Derivatives by Noncovalent Chemistry

System	Assembly Method	Properties/Applications
Biographene−HRP[35]	HRP adsorption to bovine serum albumin exfoliated graphene	~100% retention of enzyme activity
GO−BSA[36]	Hydrophobic binding	Used for efficient self-assembly of nanoparticles on GO
GO/GOx[37]	1-Pyrenebutyric acid−N-hydroxysuccinimide ester modified GO and GOx was linked by succinimidyl ester−amine nucleophilic addition	Glucose detection limit of 2 μM
GO−HRP[38]	Hydrophobic interaction with luminol-modified GO	Hydrogen peroxide detection limit of 47 fM at physiological pH
GO−lysozyme	Electrostatic interactions	Separation of lysozyme from biofluids
rGO−Aptamer[39]	Hydrophobic interactions	Picomolar detection of thrombin by fluorescence
rGO−DNA[40]	Hydrophobic interactions	Femtogram detection of cancer biomarker proteins by fluorescence
rGO−Horseradish peroxidase rGO−oxalate oxidase (OxOx)[20]	Hydrophobic interactions	• Maximum loading of HRP and OxOxwas 1.3 and 12 mg/mg, respectively • 50% reduction in HRP activity at higher reduction of GO
GO−horseradish peroxidase[41]	Multiple binding interactions	Partial loss in structure and activity of the enzyme
GO−PEG/trypsin[42]	Electrostatic binding	High thermal stability (60%−70% retention in activity at 80°C)
rGO−lipase[43]	Hydrophobic interactions	Enhanced activity of lipase (1.5−2 times) upon immobilization
GO−chymotrypsin[44]	Linked using a pyrene tripod	Active enzyme only when linker was present
GO−hemoglobin[45]	Gelation by electrostatic interactions	Peroxidase activity in methylene chloride, toluene, and chloroform
GO−tyrosinase[46]	Electrostatic and hydrophobic interaction	−0.01 nM catechol detection
GO−glucose oxidase GO−bilirubin oxidase[47]	Sol−gel synthesis	Biofuell cell was fabricated with a maximum power density of ~25 μW/cm^2 with a lifetime of 7 days
Biophilized GO−hemoglobin	Noncovalent binding of proteins to surface-passivated GO	All the enzymes preserved their secondary structure and activity on passivated surface, whereas

(Continued)

Table 8.2 Approaches for Biofunctionalization of Graphene Derivatives by Noncovalent Chemistry *Continued*

System	Assembly Method	Properties/Applications
Biophilized GO–glucose oxidase Biophilized GO–cytochrome *c* Biophilized GO–catalase Biophilized GO–horseradish peroxidase (HRP)[21]		bare GO induced significant denaturation of the enzymes (30%–80%)
GO–cationized glucose oxidase (GO*x*)[48]	Electrostatic interaction between polycation-modified enzyme and negatively charged GO	Unprecedented kinetic stability of the enzyme, where the enzyme was active for 40 days at 40°C vs. native enzyme that was active for just 6 h
GO–(GO*x* + HRP)-polyacrylic acid[49]	Protein polymer conjugates were adsorbed on GO by multiple interactions	GO-stabilized enzyme cascading under extreme pH, surfactant concentration, and/or temperature
Base washed GO– (bovine serum albumin, hemoglobin, or lysozyme)[50]	Hydrophobic interactions	Higher denaturation of the enzymes compared to as-prepared GO, which was bypassed by biophilization strategy[21]
GO–fibrin[51]	Electrostatic interactions	Tethering of osteoblast cells on GO
GO–gelatin[52]	Multiple interactions	Enhanced osteogenic differentiation of MC3T3-E1 cells and alkaline phosphatase activity than bare GO

8.4 CHARACTERIZATION OF COMPOSITES

Methods are used to determine two key aspects of the biohybrid materials. First, bioconjugation or adsorption of biomolecules onto the graphene is examined and then structure of the adsorbed biomolecule is investigated. Binding is often examined by classical equilibrium binding method, fluorescence quenching assay, zeta potential measurements, or other titrations and the binding confirmed by atomic force microscopy (AFM), scanning electron microscopy, and/or powder X-ray diffraction (XRD).[21]

The structure and biological activities of biomolecules in the biohybrid are then assessed. Circular dichroism (CD) spectroscopy is one of the best tools to analyze protein or nucleic acid secondary structures in the biohybrids but it is limited to the solution phase.[53] Infrared spectroscopy is also used very effectively to access any changes in the secondary structures of proteins in the biohybrids by monitoring the amide bands.[54] The biological activities of the biohybrids are generally examined by colorimetric or fluorescence bioassays that are specific to the system of interest. These methods are discussed below using specific examples from our laboratory as model systems.

8.4.1 EQUILIBRIUM BINDING STUDIES

The biohybrid formation via noncovalent interactions often results in an equilibrium consisting of the biohybrid, the biomolecule, and the nanosheet, at the zero order approximation. The equilibrium constant defines the relative concentrations of these species and the goal is to move the equilibrium toward the biohybrid as much as possible. In addition to the normal equilibrium treatment, one also has to take into account the presence of multiple binding sites on the nanosheet where there could be multivalent binding, cooperative binding, or multilayer binding. In this chapter, the much simpler model of Langmuir–Freundlich isotherm is applied (Eq. 8.1).[55]

$$\frac{Q}{Q_m} = \frac{K_b C^{1/n}}{1 + K_b C^{1/n}} \tag{8.1}$$

where Q is the amount of protein adsorbed per gram of GO, Q_m is the saturated adsorption, K is the LF binding constant, C is the equilibrium concentration of the biomolecule (μM), and n is the heterogeneity index, which is unity for a monolayer adsorption.

The mechanism of adsorption can be complex but there are several general steps involved in the adsorption process, including (1) diffusion of the biomolecule to the surface of the nanosheet, (2) binding of the biomolecule to one or more binding sites on the nanosheet surface, (3) relaxation of the bound biomolecule at the nanosheet surface, often resulting in some loss in its secondary/tertiary structure, (4) dissociation of the biomolecule from the surface, and (5) diffusion of the released biomolecule into the bulk. The Langmuir–Freundlich isotherm takes these steps into account. In addition, any cooperative behavior is also taken into account. That is, binding of one biomolecule next to another at the surface may be promoted or inhibited and this can lead to enhancement or inhibition of binding. A basic assumption that all binding sites are available to the biomolecule at all levels of loading is made, but as the surface gets filled it becomes increasingly difficult to occupy free sites due to entropic reasons. Thus, the binding model described here is rudimentary but adequate to illustrate characterization of these biohybrids.

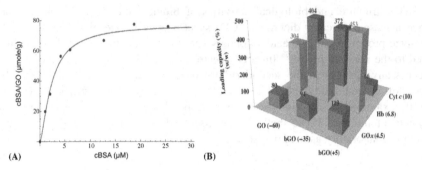

FIGURE 8.3

(A) Langmuir–Freundlich isotherm of cBSA binding to GO. Binding parameters were calculated by nonlinear fit ($R^2 = 0.997$) of the isotherms following Eq. 8.1. (B) Maximal loadings (%w/w) of glucose oxidase (GOx), hemoglobin (Hb), and cytochrome c (Cyt c) to biophilized graphene oxide sorted by zeta potential charge in the parenthesis GO(−60), bGO(−35), bGO(+5), at pH 7 (© American Chemical Society).[21]

Equilibrium binding studies are performed to find the extent of adsorption of the biomolecule onto the nanosheets and the binding constant estimated from the isotherms using best fits to the chosen binding model. The extent of adsorption is quantified as a function of biomolecular concentration, at fixed substrate concentrations. Fig. 8.3A shows the binding isotherm for chemically modified bovine serum albumin (cationized BSA or cBSA) to GO. The protein has been chemically modified by amidating carboxyl groups with a polyamine such that the net charge on the protein is controlled as desired. The chemical modification did not alter the structure of the protein to a significant extent but did distort its overall structure to some extent, as evidenced by CD spectroscopy. The data were fitted to Eq. 8.1 and best fits to the data indicated a binding constant (K_b) of 2.73 × 10^5 M^{-1} and surface heterogeneity index (n) of 0.67.[21]

Alternatively, saturation adsorption of the protein onto the substrate is examined at high protein concentrations and the values compared among several different proteins and nanosheets. Such studies are suitable when a large set of proteins and binding surfaces are under evaluation. Also, the binding isotherms require a substantial amount of protein, whereas the saturation binding studies are single-point measurements but the latter do not provide affinity constants or other binding parameters. To find the maximal loading, it is important to confirm that the protein concentration is above the saturation limit or that there is excess free protein in the solution, after equilibration of the system. In our lab, we choose two or three extreme concentrations to ensure that the binding has been saturated. Fig. 8.3B shows the maximal loading capacity (w/w) for three different proteins bound to GO and GO/cBSA biohybrids. Using the charge-modified BSA, two biohybrids bGO(−35) and bGO(+5) with net charges of −35 and +5 are produced and binding of three different proteins to the three kinds of GO nanosheets with increasing charge have been examined. The data clearly indicate the inter-twined connection between the charge on the nanosheet versus the charge on the biomolecule. In the case of cytochrome c (Cyt c), which is positively charged at pH 7, protein binding is directly related to the net charge on the nanosheet, while in the case of hemoglobin (Hb), which is slightly negatively charged, the binding increased only slightly with net charge on the nanosheet. The binding of the strongly negatively charged glucose oxidase (GOx) was the lowest and did not vary much with charge on the biohybrid. Thus, electrostatic interactions play an important role in controlling protein binding to the biohybrids but they are only a part of the sum of interactions. Further work is needed to understand the details of these interactions and their magnitudes. In addition to the saturation binding studies, one can perform the fluorescence quenching assays to further examine the equilibrium binding model.

8.4.2 FLUORESCENCE SPECTROSCOPY

The low-lying electronic states of graphene provide a convenient mechanism to quench the fluorescence from a number of different fluorophores and this is explored to estimate binding affinities of proteins.[56,57] For example, when the protein is tagged with a highly fluorescent reagent, such as fluorescein isothiocya-nate (FITC), the binding of the resulting FITC-BSA to GO can be examined by fluorescence spectroscopy. Upon successful adsorption of BSA-FITC to GO, its fluorescence is quenched (Fig. 8.4A).[57] The extent of quenching was monitored as a function of GO concentration and the data fitted to the Stern−Volmer equa-tion that has been suitably modified (Eq. 8.2).[58]

$$\frac{F_0}{F_0 - F} = \frac{1}{f_a K_a [Q]} + \frac{1}{f_a} \tag{8.2}$$

In the above equation, $[Q]$ = concentration of the quencher; F_0 = fluores-cence intensity when $[Q] = 0$; F = fluorescence intensity at given $[Q]$; K_a = bimolecular quenching constant; and f_a = maximal fraction of the initial fluores-cence accessible to the quencher.

The quenchings of fluorescence from cBSA and BSA upon binding to GO were compared in Fig. 8.4B and C. Evidently, the higher affinity of cBSA resulted in greater quenching of the fluorescence by GO and poor quenching by BSA was due to its weaker affinity to GO. The magnitude of this difference on interaction was about an 80-fold increase in the Stern−Volmer constant (Fig. 8.4D). Affinity constant of cBSA to GO from Stern−Volmer analysis was estimated as 228 mL/mg and that of BS3A was about 3 mL/mg. Fluorescence experiments offer quantitative binding estimates, while requiring only small amounts of protein and substrate. The weaknesses of this technique are the chemi-cal tagging step which requires reactive amino or carboxylic groups on protein so that its binding to the substrate can be tracked. False-positive results due to scat-tering and light absorption by GO also need to be accounted for. We recommend

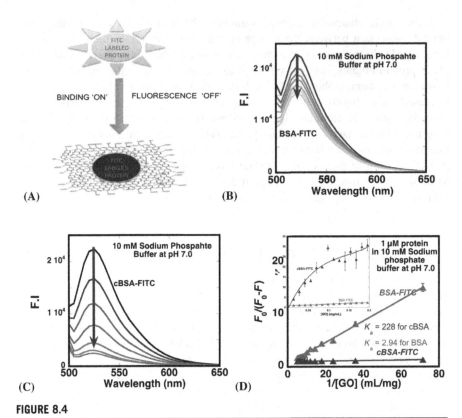

FIGURE 8.4

(A) Quenching of cBSA-FITC fluorescence by GO. Fluorescence spectra show the quenching of fluorescence of cBSA-FITC (B) and BSA-FITC (C) by GO. Clearly, quenching is very prominent in the case of chemically modified BSA. (D) Modified Stern–Volmer plots (Eq. 8.2) for the quenching of cBSA-FITC (blue lines) and BSA-FITC (red lines) by GO. The inset shows the Stern–Volmer plots (© American Chemical Society).[21]

choosing low GO concentrations to reduce scattering as well as light absorption by GO. Some fluorescent molecules may influence the binding event ("label matters")[59] so control experiments are required to account for the direct affinity of the dye molecule for the nanosheet.

8.4.3 ZETA POTENTIAL STUDIES

Zeta potential measurements are widely used for studying the colloidal stability and surface modifications in graphene and GO research.[60–63] Contributions of electrostatic interactions, and the role of solvent and ions on protein binding to nanosheets are also studied using zeta potential measurements. Binding of proteins to 2-D layers of α-zirconium phosphate was monitored using this technique.[64]

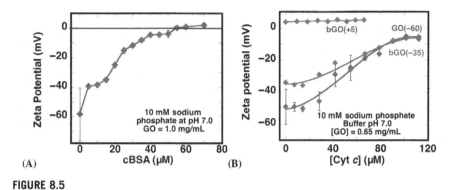

FIGURE 8.5

(A) Zeta potential of GO gradually changes from −60 mV to +5 mV upon addition of cBSA. (B) Zeta potential as a function of Cyt *c* loading on bGO. Protein binding increased the zeta potential in all cases (© American Chemical Society).[21]

Binding of biomolecules to negatively charged GO surface (due to the ionizable oxygen groups present on GO) results in an increase or decrease in magnitude of the net charge of the suspension. Gradual addition of positively charged cBSA (0−70 μM) to a suspension of the nanosheets resulted in a change of the net charge of GO from −60 to +5 mV at pH 7.0 (Fig. 8.5A). Binding of positively charged Cyt *c* to GO with various surface groups also causes a change in zeta potential. These changes support the role of the electrostatic interactions in the binding mechanism. Zeta potential measurements often require large samples, but it is a nondestructive method and the sample can be recovered after the analysis.

8.4.4 ATOMIC FORCE MICROSCOPY

AFM images are often used to distinguish the biomolecules bound to the surface of the nanosheets from those bound at the edges. Solution-phase AFM imaging is preferred to study the topography of the bound biohybrid, where artifacts due to sample drying can be avoided. Furthermore, quantitative analysis by AFM height traces are used to distinguish monolayer vs. multilayer adsorption of biomolecules. Orientation of biomolecules on surfaces can be examined by AFM height traces. AFM images of cBSA adsorbed GO (collected in solution over mica sheets) shows the presence of protein islands on the GO surface, compared to the bare GO surface (Fig. 8.6A). The height trace across the sheet shows about a 3−4-nm step, which is closer to the dimensions of cBSA (prolate ellipsoid 140 × 40 × 40 Å).[65] Similarly, the hybrids of Cyt *c*, Hb, and GO*x* to GO and cBSA-coated GO (bGO) are analyzed using this methodology (Fig. 8.6B). The height profiles of these images matched well with the theoretical size of the protein molecule. Microscopy data of the biohybrids and their analyses can often be verified by another method, the powder XRD, which is an orthogonal technique to examine nanosheet structures and hybrids.

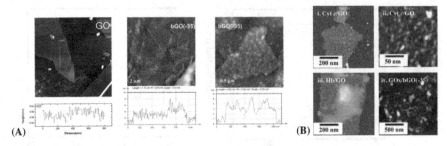

FIGURE 8.6

(A) AFM images of GO and cBSA adsorbed at different magnifications imaged using a liquid cell over mica. cBSA decoration increased the plate thickness. (B) Binding of Cyt *c* and Hb to GO and the binding of GO*x* to bGO(−35) is clearly demonstrated in these images (© American Chemical Society).[21]

8.4.5 X-RAY DIFFRACTION

Powder XRD is a powerful tool to study the interaction of layered materials with proteins and DNA.[64,66,67] Slow evaporation of solvent from suspensions of protein/layered biohybrids triggers the assembly of proteins and nanolayers into stacks with certain spacings between the layers (Fig. 8.7A). These layers can diffract the X-rays, and the corresponding d-spacings obtained from the diffraction patterns often correspond to the diameter of the biomolecule trapped between the plates (Fig. 8.7B). The layered morphology of four different proteins and GO was proved by this technique where the theoretical diameters matched with the observed d-spacings of the biohybrid films (Fig. 8.7B). Wide-angle XRD machines are often not suitable to analyze biohybrid assemblies of large proteins (>80 Å) and nanosheets but small-angle X-ray spectroscopy (SAXS) is suitable for measuring these large interlayer spacings.

8.4.6 CD SPECTROSCOPY

When the biohybrids are formed, it is often important to examine the secondary structure of the bound biomolecule. This is because the loss of solvent and change in the microenvironment around the biomolecule due to the formation of the biohybrid can distort its three-dimensional structure. However, the native structure of the biomolecule is important to retain its biological function. The interactions at the bio−nano interface often result in significant loss of biomolecular structure, a fundamental problem.[68] Biomolecular structural changes are followed conveniently by comparing the CD spectra of the biomolecules before and after the biohybrid formation. The magnitude of structural deformation of the bound biomolecule, e.g., depends on the interactions between the functional groups on the protein and those on the GO surface. These interactions are not well-understood, they are complex, could be additive, cooperative, and various reports

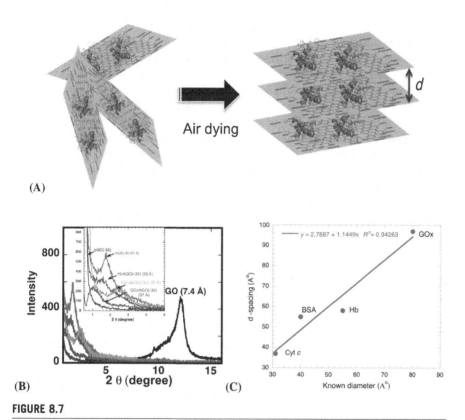

FIGURE 8.7

(A) Protein-bound, exfoliated GO sheets in solution (left), which are reassembled to a locally ordered layered structure (right) upon air-drying the suspension on a glass slide. (B) Powder XRD data of protein/bGO(−35) films. (C) The d-spacings correlated well with the known diameters of the corresponding proteins bound to the bGO nanosheets (© American Chemical Society).[21]

suggest that they are primarily driven by hydrophobic and/or electrostatic forces.[41,68,69] To quantitate the changes in the secondary structure of the biomolecule, CD signal intensity at 222 nm of the bound protein is compared with that of the corresponding unbound protein, following Eq. 8.3. The extent of retention of the structure of the bound protein (%) was calculated by comparing the α-helicity of the unbound protein with that of the bound protein.

$$\alpha - \text{Helicity retained} = \left(\frac{\dfrac{\theta_{222,b}}{b}}{\dfrac{\theta_{222,f}}{f}} \right) 100 \tag{8.3}$$

where $\theta_{222,b}$ and $\theta_{222,f}$ are the observed ellipticities of the bound and free protein, respectively, and b and f are the concentrations of the bound and free proteins, respectively. Percentage α-helicity retained upon biohybrid formation was calculated by comparing it with those of the corresponding free proteins. One important consideration is to avoid artifacts in the CD spectra of suspensions, which are prone to scatter the incident radiation and potentially distort the observed CD spectra. Short path lengths and low concentrations are preferred to minimize scattering of the incident radiation in the experiment. In addition, we also strongly encourage testing the influence of the nanosheet suspensions on the CD spectra. This is done by doing the tandem cell experiment, where two separate cells are loaded with the nanosheet suspension and the protein solution. Both cells are placed in series in the spectrometer and the influence of the nanosuspension on the CD spectrum of the biomolecule examined. When the path lengths are kept low (0.05 cm) and biomolecule absorbance around 222 nm is kept under 0.25, these artifacts can be reduced considerably. The concentration of the nanosheets and their size also strongly influences the scattering, and hence, these experiments are to be done very carefully.

CD spectra of hemoglobin bound to GO indicated significant loss of its structure (Fig. 8.8A). Similar observations are made for seven other proteins and the percent structure retention ranged from 5% to 70% (Fig. 8.8B). This analysis points out that the GO surface is not suitable for protein structure retention but our laboratory developed appropriate surface modification methods to passivate the GO surface. GO passivated with cationized BSA coating (bGO(+5)) improved the extent of structure retention (Fig. 8.8B) and a nearly native-like structure has been noted for a number of proteins bound to bGO(+5). To confirm the findings in CD spectroscopy, biological activity assays should be performed because any change in structure of a biomolecule could affect its function.

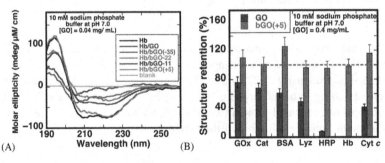

FIGURE 8.8

(A) CD spectra of hemoglobin bound to a variety of graphene surfaces. Strong distortion of the protein structure was noticed when bound to GO (see the legend). (B) Extent of structure retention of seven different proteins at GO and cBSA-coated GO (bGO), calculated using Eq. 8.3 and from the CD data in the left panel (© American Chemical Society).[21]

8.4.7 **BIOLOGICAL ACTIVITY STUDIES**

Applications of the biohybrids require that the biological function of biomolecules adsorbed on graphene be preserved to a substantial extent. Biocatalysts, e.g., should have sufficient activity for their use as catalysts in the laboratory or industry. Enzymatic activities are measured by standard assays, with the characteristics compared before and after binding to graphene (Fig. 8.9). The activities are recorded relative to the activity of the free enzyme in the absence of graphene but under otherwise identical conditions of pH, buffer, and temperature. Usually, lower activities are noted when these are bound to GO and this could be due to a number of reasons which include reduced access to the active site of the bound enzyme, decreased diffusion to the interior of the biohybrid, loss in the active site structure in the biohybrid and other factors. Activities of some enzymes bound to GO and bGO($+5$) are compared in Fig. 8.9. All the enzymes bound to bGO($+5$) examined in our laboratories showed native-like activities, whereas enzymes bound to GO showed significant reductions. Thus, the preloaded cBSA on the GO surface must have blocked certain hot spots on GO that are likely to denature the bound enzymes. Therefore, bGO($+5$) is proving to be an excellent alternative to GO for the adsorption of enzymes such as these.

8.5 **BIOCATALYSIS USING GRAPHENE PLATFORM**

Graphene and GO are good platforms as biocatalyic supports, owing to their high surface area, light weight, chemical/thermal stability, low cost, high strength, and attractive electronic properties.[41] Despite the partial denaturation and varying extents of inhibition of enzyme activities, graphene derivatives provide good platforms for biocatalysis applications, especially under nonphysiological environments. Here, three different routes to improve the biocatalytic performance of enzymes bound to GO are discussed.

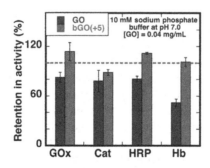

FIGURE 8.9

Enzymatic activity of enzymes of graphene and cationized BSA coated GO (bGO ($+5$)) with respect to those of the corresponding free enzymes (© American Chemical Society).[21]

8.5.1 PRESERVATION OF STRUCTURE AND FUNCTION OF BOUND ENZYMES

Conservation of biological structures is vital for its specificity and efficiency in function. Protein conformational changes are often encountered upon binding to a solid surface and they are attributed to the opposing contributions of the surface energy of the solid and the free energy of denaturation (ΔG_{D-N}) of the protein. Specifically, proteins with low denaturation free energies (less positive values) unfolded or denatured while consuming surface energy of the solid surface. The surface free energy of GO was estimated to be around 62.1 mJ/m.[70,71] Even though the large number of functional groups and the large surface area of GO favor effective binding of proteins, the hydrophobic nature of the surface is not benign to many proteins.[72] Proteins with poor stabilities or those that can interact with the hydrophobic regions of GO tend to lose their structure more readily,[69] as illustrated by HRP/GO.[18] Therefore, converting GO into a biocompatible material is both challenging and rewarding. Our attempt has been to control the surface characteristics of GO in a systematic manner, so that the structure and function of bound biomolecules can be improved to attain those of the corresponding unbound biomolecules.

The hypothesis is that by masking the hot spots of GO by a biocompatible molecule, the sites that contribute unfavorable interactions with enzymes can be passivated. The biological coating should have a high affinity for these spots on GO, should be strongly hydrophilic to support extensive hydration sites, and support binding of enzymes and other biomolecules on the modified GO. Since GO is strongly negatively charged, we surmised that positively charged soft proteins will be most suitable for GO modification. BSA has been chemically modified by attaching polyamine groups to its surface such that the chemically modified BSA is positively charged (cBSA) and cBSA possess high electrostatic affinity for the negatively charged GO. Unmodified BSA, however, is unsuitable as it is negatively charged at neutral pH, and it indicated only a very poor affinity for the anionic GO.

Using the above strategy, we have prepared cBSA-loaded GO, called bGO, with a net positive charge of +5 units (bGO(+5)). Seven different proteins and enzymes were tested for binding and activity retention with bGO(+5). Cyt c bound to bare GO lost almost all of its CD signal at 222 nm but the CD signal at 222 nm improved with increased cBSA loading on GO and recovered fully on bGO(+5). The retention of the activity and structure of the protein, therefore, was proportional to the extent of cBSA modification on GO and there was an optimal loading of cBSA that provided the best structure retention. For example, a plot of the extent of structure retention vs. the charge on GO that has been modified with cBSA to increasing extents, as monitored by the zeta potential, indicated continuous improvement in bound enzyme structure retention with increased loading of cBSA or increased charge on the passivated GO (Fig. 8.10A).

Activities of the bound enzymes proportionately depended on the extents of their structure retention after binding to GO, with one exception. Cyt *c*/GO showed exceptionally high activity, 15-fold greater than Cyt *c* (Fig. 8.10B) despite the fact that it lost most of its structure. Similar results were noted when Cyt *c* was loaded onto the GO-modified electrode and the electrochemical activity was enhanced dramatically. Denaturation of the protein secondary structure and exposure of the heme from Cyt *c* onto GO was suggested to be responsible for the enhanced electrochemical signals of Cyt *c*.[73] We did not observe similar enhancements in the activities for other proteins and enzymes that contained hydrophobic prosthetic groups, and therefore it is not a general phenomenon that GO would enhance the activities of enzymes. On the contrary, most enzymes we examined lost some activity on binding to GO.

To further test the above hypothesis that an increase in cBSA loading on GO had a positive influence on the bound enzyme/protein structure, we examined if there has been any correlation between the percent activities of the bound enzymes and their corresponding percent helicity noted in their CD spectra (Fig. 8.11). The plot of 10 different data points clearly indicated a linear correlation between the structure retention and activity. Enzymes/proteins bound to hydrophobic regions or hot spots of the GO sheets tend to lose some of their secondary structure and this loss in structure results in lower activities. As these regions are increasingly covered by cBSA, which functioned as a sacrificial protein layer, the bound enzyme is not exposed to these hot spots and the fraction of active enzyme/protein increased proportionately. Increased fractions of the enzyme molecules bound to passivated regions of the modified GO or bGO are thought to be responsible for improved activities. Thus, it is likely that certain

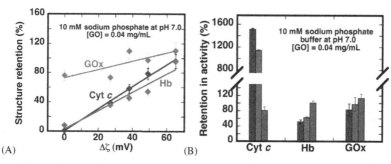

(A)   (B)

FIGURE 8.10

(A) Retention of structure increase with biophilization of GO. Increase in zeta potential because of biophilization is used as a proxy to represent GO modification. (B) Activity of proteins on GO (*left side bars*), partially biophilized surface (*middle bars*), and fully biophilized GO (*right side bars*) show native-like structure and activity of the enzymes after passivation of GO (© American Chemical Society).[21]

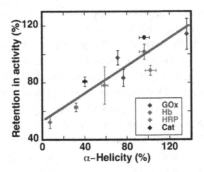

FIGURE 8.11

Correlation of activity retention with α-helicity (%) of four distinct proteins bound to GO preloaded with increased concentrations of cBSA. The linear relationship between structure and activity confirms that improved structure translated to improved activities (© American Chemical Society).[21]

regions on GO are not conducive for protein structure retention but upon passivation of these regions with cBSA, enzyme binding promotes the retention of native-like structures of the bound enzymes and activities of the bound enzymes are preserved to a larger extent.

Enzyme loading onto GO also improved after coating with cBSA, in most cases. Thus, modification of GO with cBSA, preloading the surface, is a step forward for the development of functionally active graphene biomaterials or biohybrids with superior properties when compared to untreated GO. Our next goal has been to improve enzyme thermal stability in the biohybrids.

8.5.2 THERMODYNAMIC STABILITIES OF ENZYMES

One major challenge in the use of enzymes in the laboratory or in industry is their poor stability under ambient, nonbiological conditions. Our group has been working for decades on improving enzyme stability by encasing them in layered inorganic materials with considerable success.[15,74] Thermodynamic stabilities of enzymes, e.g., can be improved by confining then in a 2-D matrix.[75] Enzyme stabilization under nonphysiological conditions is beneficial for their laboratory and industrial use. Graphene proved to be an excellent matrix for the stabilization of various enzymes against heat and chemical denaturants. For example, protease activity of trypsin was improved 70% at 70°C by embedding in PEG-modified GO.[76] Preservation of peroxidase activity of hemoglobin in organic solvents was preserved by hydrogel formation with GO.[45] A bionanocatalyst of glucose oxidase (GO*x*) and horseradish peroxidase (HRP) embedded in poly(acrylic acid) (PAA) and GO indicated high activity (Fig. 8.12).[77] By combining the conjugation of proteins with a synthetic polymer with biohybrid formation with GO improved protein stability,

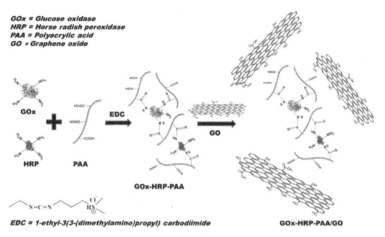

FIGURE 8.12

Synthesis of GOx-HRP-PAA/GO bionanocatalyst. PAA is represented as thick lines with -COOH groups and GO is represented as a sheet. Carbodiimide chemistry is used to crosslink amine groups on the enzymes to COOH groups in PAA (© American Chemical Society).[49]

and the embedded proteins survived multiple stresses of temperature (65°C), pH (2.5−7.4) and exposure to surfactants (4 mM sodium dodecyl sulfate (SDS)).

The above catalyst showed 100% activity at 65°C, where proteins normally begin to denature, and the unmodified enzymes lost 80% of their enzymatic activities (Fig. 8.13A).[42] At the extreme pH of 2.5, the unprotected enzyme is completely deactivated while the enzyme-PAA−GO hybrid retained 100% of its activity (Fig. 8.13B). A similar stabilizing effect was observed for the biohybrid when it was exposed to 4 mM SDS, at pH 2.5 (Fig. 8.13B). The stabilization effects of GO matrix expand the practical applications of biocatalysts under biologically challenging conditions such as high temperatures, low pHs, and high surfactant concentrations.

8.5.3 KINETIC STABILITY OF ENZYMES

The kinetic/storage stability of an enzyme is another important metric.[78] Catalytic activity of the enzyme after multiple catalytic cycles or after prolonged use or storage is essential for their use in a bioreactor or a biofuel cell. Controlled electrostatic interactions between GOx and GO was used to prepare "stable-on-the-table" glucose oxidase (GOx) biohybrids (GO/GOx).[48] Catalytic activities of GO/GOx biohybrids at specific time intervals after storage at 40°C are compared with the activity of freshly made solution of GOx. At the best, the GO/GOx hybrid

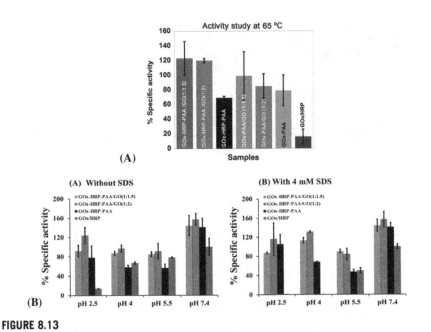

FIGURE 8.13

(A) Enzymatic activities of the protein-PAA–GO hybrids at high temperature. (B) Activity of the biohybrid in the presence and absence of surfactant over a wide pH range (© American Chemical Society).[49]

showed a half-life of 40 days at elevated temperatures (40°C), whereas the unprotected enzyme had a half-life of just 6 hours (Fig. 8.14A).[48]

The improved stability of GO/GOx could be due to several reasons, including increased kinetic barrier for denaturation, prevention of aggregation of the enzyme, and protection against attack by proteases during storage.[48] Since the enzymes are adsorbed on the GO sheets, the chances of aggregation and aggregation-induced inactivation of the enzyme are inhibited to a significant extent. GO is known to be an inhibitor for some proteases, which also helps in reducing the protease-mediated degradation of the enzyme during storage because proteases are ubiquitous in the environment and can degrade most proteins over short periods of time.[69] Along these lines, the radical scavenger effect of GO protected horseradish peroxidase from free radical mediated denaturation for up to 60 hours, which is 10 times higher than that of the unprotected enzyme.[79] This radical scavenging property of GO could be protecting GOx from H_2O_2, the byproduct of glucose oxidation by oxygen, during its catalytic function.[80] The activation energy (E_a) for the deactivation pathway of the biohybrid was calculated from the kinetic data (Fig. 8.14B). The activation energies correlated strongly with the charge on GOx. This increased kinetic barrier by charge tuning followed by adsorption onto GO is responsible for the extraordinary kinetic stability of the GO/GOx biohybrid reported above.[48]

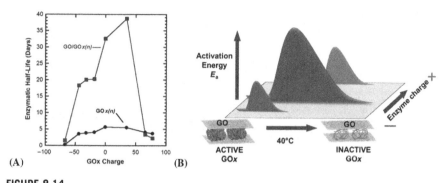

FIGURE 8.14

Half-life of GO*x* bound to GO in comparison with that of the free enzyme. Charge of GO*x* was modified by controlled amidation of its carboxyl group with polyamines. (B) Suggested mechanism of kinetic stability of the enzyme by forming GO hybrid (© American Chemical Society).[48]

8.6 BIOSENSING

Detection of biomolecules with the graphene platform is growing in the biomedical field.[81] Fluorescence-based method sensors exploit the fluorescence quenching behavior of graphene toward many fluorophores, whereas the electrochemical sensors harness the electroactivity and conductivity of graphene.

8.6.1 FLUORESCENCE-BASED GRAPHENE BIOSENSORS

Quenching of the emission from fluorophores by graphene is often used to fabricate biosensor systems.[82,83] In one approach, the analyte has a higher affinity for the fluorophore than graphene, and the fluorophore is dissociated from graphene and captured by the analyte when the target molecule is present (Fig. 8.15, Route A).[84] Single-stranded DNA was detected using this approach, where a fluorescently labeled complementary strand was quenched on graphene and the addition of the target strand results in an increase in fluorescence after double helix formation. It has been noted that the double helical DNA has poor affinity to GO while the single-stranded DNA binds well.[85] Adenosine sensors developed using this strategy show a detection limit as low as 20 pM.[83] In another approach, competitive binding of the analyte to graphene loaded with a fluorophore can result in the displacement of the fluorophore from graphene and recover the fluorescence arising from the free fluorophore in the solution.[69] Thus, the design of these "turn-on" florescence assays is attractive for analytical assays (Fig. 8.15, Route B). Highly sensitive protein sensors are designed by this approach for the detection of eight different proteins including hemoglobin, lysozyme, BSA, and myoglobin,

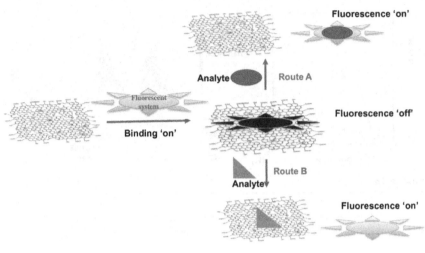

FIGURE 8.15

This scheme illustrates two approaches for biosensing on graphene using fluorescent systems.[69,84]

achieving 10 nM detection limits.[69] Likewise, the electrochemical activity of graphene can also be used to fabricate highly sensitive biosensors.

8.6.2 ELECTROCHEMICAL GRAPHENE SENSORS

Biosensors are fabricated using graphene derivatives to detect analytes such as glucose, hydrogen peroxide, nicotinamide adenine dinucleotide, thrombin, DNA, and so on.[85–87] The high surface area, good electrochemical activity, easy fabrication, and inexpensive graphene preparations provide ample opportunities for rapid detection of these target systems when coupled with the specificity of biomolecules for detection. A large number of studies in this field are devoted to the detection of glucose because of its biomedical significance. These sensors show detection limits for glucose ranging from 0.1 μM to 1 mM, even in whole blood, which is far below normal blood glucose levels (4−6 mM).[84,87] Other examples of glucose biosensors are listed in Tables 8.1 and 8.2. Hydrogen peroxide is detected using the peroxidase activity of HRP, hemoglobin, or other heme-containing enzymes with high specificity. The detection limits of these systems range from 21 nM to 5 μM depending on the approach and methodology implemented.[87]

The major challenge in this field is the detection of real samples in biological media with naturally interfering substances at biologically relevant levels. Commercially available glucose-detecting devices are becoming cheaper and more widely available, but methods for continuous monitoring of glucose levels in live systems is a current, unmet challenge. Also, the detection of more challenging analytes other than glucose or hydrogen peroxide may have a major impact from both fundamental science, biomedical applications, and commercial points of view.

8.7 DRUG DELIVERY APPLICATIONS

A large surface area of the two-dimensional sheets per unit mass is an attractive metric for the design of efficient and multivalent drug-delivery vehicles, where the flat surfaces allow for maximum loading of drug per unit area as well as the attachment of specific biological recognition elements for targeted drug delivery and a means to image the fate of the drug in the biological system. GO with its water dispersibility and hydrophobic patches is a good candidate for the distribution of sparingly water-soluble hydrophobic drug molecules.[3] The $\pi-\pi$ stacking and electrostatic interactions between the hydrophobic drug molecules are used to deliver various anticancer drugs as well as specific segments of nucleic acids in both in vitro and in vivo systems with graphene as a high-capacity carrier. For example, PEG-modified GO was used to deliver anticancer drug SN38 (camptothecin (CPT) analogue) to colon cancer cells (HCT 116), where the composite was 1000 times more potent than the free drug, which is poorly soluble in water.[88]

Stimuli-responsive drug-delivery systems are also being fabricated with graphene media, where the drugs are released in response to pH, light, glutathione concentration or other stimulants.[89-91] These drug-delivery systems, coupled with photo-responsive systems, can be used for multifunctional drug release.[92] Near-infrared absorption of GO was used for tumor ablation in mice with 100% potency after 48 hours of exposure.[93] These systems have high potential because currently used chemotherapy drugs attack nonmalignant cells causing severe side effects in patients. Examples of these systems are listed in Tables 8.1 and 8.2. The toxicity effect from graphene derivatives could be a roadblock for these systems, especially when the enzymatic degradation of the sheets could release mutagenic substances, which can be triggered by light.[92] Detailed analysis of biological clearance and side effects of these systems will have to be investigated prior to their implementation for therapy, which may take years to decades.

8.8 CONCLUSIONS

Graphene as a matrix for various applications in biological systems has been discussed in this chapter. The application of graphene in several biological fields is amazing in that no other material has found such usefulness in so many different fields. Graphene's physical, chemical, mechanical, and electronic properties are harnessed in a biological context for the fabrication of biocatalysts, sensors, drug-delivery systems, and so on. Graphene derivatives impart stability, durability, and sustainability to biohybrids for prolonged and optimum function. In our laboratory, properly tuned GO/enzyme hybrids are much more stable when exposed to heat, chemical denaturants, or ambient storage over time, when compared to unprotected enzymes. Biosensors coupled with graphene derivatives were able to detect analytes at low picomolar concentrations, which could be useful for early diagnosis of various diseases. Partial hydrophobicity of GO was exploited for the delivery of hydrophobic drugs, improving the bioavailability of drugs, or for improving the effectiveness of the treatment. Photosensitivity of graphene is also being used for localized delivery of drugs, which may help in minimizing the adverse side effects that are common to several drugs that are currently on the market.

A common problem in the biological application of many nanomaterials is the formation of protein corona around the materials when exposed to serum proteins, which affects the performance. Protein corona can contribute to a significant increase in the size and could interfere with or modify the surface chemistry of the nanomaterial. Studies suggest that the presence of protein corona can mask the intrinsic toxicity of the material by reducing the direct exposure of the materials to biological molecules or surfaces.[94] However, protein corona severely affects the specificity and targeting of the nanomaterials.[95] The field of nanomedicine, where nanomaterials play a key role, faces a big challenge due to the nonspecific interaction of most nanoparticles with biological molecules and surfaces. A new study shows that only 0.7% of the total nanomaterial loaded with drug molecules exposed to an in vivo system has been delivered to the targeted region, estimated from an average of 117 independent experiments.[96] The remaining 99.3% of the particles accumulated in other parts of the body; usually the spleen, liver, or kidney and potentially these can cause damage and toxicity. Thus, biofunctionalization of nanoparticles to control their interactions and distributions in various destinations could be critical. Graphene synthesized and stabilized in serum,[97] therefore, may be attractive for nanomedicine and other applications in biological fields.

REFERENCES

1. (a) Oliveira, O. N.; Iost, R. M.; Siqueira, J. R.; Crespilho, F. N.; Caseli, L. Nanomaterials for Diagnosis: Challenges and Applications in Smart Devices Based on Molecular Recognition. *J. Appl. Mater. Interfaces* **2014**. (b) Chung, C.; Kim, Y.-K.; Shin, D.; Ryoo, S.-R.; Hong, B. H.; Min, D.-H. Biomedical Applications of Graphene and Graphene Oxide. *Acc. Chem. Res.* **2013**, *46*, 2211–2224. (c) Riehemann, K.; Schneider, S. W.; Luger, T. A.; Godin, B.; Ferrari, M.; Fuchs, H. Nanomedicine— Challenge and Perspectives. *Angew. Chem. Int. Ed.* **2009**, *48*, 872–897. (d) Yang, K.; Feng, L.; Shi, X.; Liu, Z. Nano-Graphene in Biomedicine: Theranostic Applications. *Chem. Soc. Rev.* **2013**, *42* (2), 530–547.

2. Novoselov, K. S.; Geim, A. K.; Morozov, S. V.; Jiang, D.; Zhang, Y.; Dubonos, S. V.; Grigorieva, I. V.; Firsov, A. A. Electric Field Effect in Atomically Thin Carbon Films. *Science* **2004,** *306,* 666–669.

3. Shi, S.; Chen, F.; Ehlerding, E. B.; Cai, W. Surface Engineering of Graphene-Based Nanomaterials for Biomedical Applications. *Bioconjug. Chem.* **2014,** *25* (9), 1609–1619.

4. Yang, K.; Feng, L.; Shi, X.; Liu, Z. Nano-Graphene in Biomedicine: Theranostic Applications. *Chem. Soc. Rev.* **2013,** *42* (2), 530–547.

5. Feng, L.; Liu, Z. Graphene in Biomedicine: Opportunities and Challenges. *Nanomedicine* **2011,** *6* (2), 317–324.

6. Bitounis, D.; Ali-Boucetta, H.; Hong, B. H.; Min, D. H.; Kostarelos, K. Prospects and Challenges of Graphene in Biomedical Applications. *Adv. Mater.* **2013,** *25* (16), 2258–2268.

7. Kim, J.; Park, S.-J.; Min, D.-H. Emerging Approaches for Graphene Oxide Biosensor. *Anal. Chem.* **2016,** *89* (1), 232–248.

8. Shen, H.; Zhang, L.; Liu, M.; Zhang, Z. Biomedical Applications of Graphene. *Theranostics* **2012,** *2* (3), 283–294.

9. Zhang, Y.; Nayak, T. R.; Hong, H.; Cai, W. Graphene: A Versatile Nanoplatform for Biomedical Applications. *Nanoscale* **2012,** *4* (13), 3833–3842.

10. Si, Y.; Samulski, E. T. Synthesis of Water Soluble Graphene. *Nano Lett.* **2008,** *8,* 1679–1682.

11. Dreyer, D. R.; Park, S.; Bielawski, C. W.; Ruoff, R. S. The Chemistry of Graphene Oxide. *Chem. Soc. Rev.* **2010,** *39,* 228–240.

12. Hsing, I. M.; Xu, Y.; Zhao, W. Micro- and Nano- Magnetic Particles for Applications in Biosensing. *Electroanalysis* **2007,** *19* (7-8), 755–768.

13. Dai, Z.; Kawde, A.-N.; Xiang, Y.; La Belle, J. T.; Gerlach, J.; Bhavanandan, V. P.; Joshi, L.; Wang, J. Nanoparticle-Based Sensing of Glycan – Lectin Interactions. *J. Am. Chem. Soc.* **2006,** *128,* 10018–10019.

14. Wang, S.; Mamedova, N.; Kotov, N. A.; Chen, W.; Studer, J. Antigen/Antibody Immunocomplex from CdTe Nanoparticle Bioconjugates. *Nano Lett.* **2002,** *2,* 817–822.

15. Deshapriya, I. K.; Kumar, C. V. Nanobio Interfaces: Charge Control of Enzyme/Inorganic Interfaces for Advanced Biocatalysis. *Langmuir* **2013,** *29* (46), 14001–14016.

16. Nel, A. E.; Madler, L.; Velegol, D.; Xia, T.; Hoek, E. M. V.; Somasundaran, P.; Klaessig, F.; Castranova, V.; Thompson, M. Understanding Biophysicochemical Interactions at the Nano-Bio Interface. *Nat. Mater.* **2009,** *8,* 543–557.

17. Pastré, D.; Piétrement, O.; Fusil, S.; Landousy, F.; Jeusset, J.; David, M.-O.; Hamon, L.; Le Cam, E.; Zozime, A. Adsorption of DNA to Mica Mediated by Divalent Counterions: A Theoretical and Experimental Study. *Biophys. J.* **2003,** *85,* 2507–2518.

18. Nepal, D.; Geckeler, K. E. *Interactions of Carbon Nanotubes with Biomolecules: Advances and Challenges. Advanced Nanomaterials;* Wiley-VCH Verlag GmbH & Co. KGaA: Weinheim, **2010,** 715–742.

19. Paul, T.; Bera, S. C.; Agnihotri, N.; Mishra, P. P. Single-Molecule FRET Studies of the Hybridization Mechanism during Noncovalent Adsorption and Desorption of DNA on Graphene Oxide. *J. Phys. Chem. B* **2016,** *120* (45), 11628–11636.

20. Zhang, Y.; Zhang, J.; Huang, X.; Zhou, X.; Wu, H.; Guo, S. Assembly of Graphene Oxide–Enzyme Conjugates through Hydrophobic Interaction. *Small* **2012,** *8* (1), 154–159.

21. Pattammattel, A.; Puglia, M.; Chakraborty, S.; Deshapriya, I. K.; Dutta, P. K.; Kumar, C. V. Tuning the Activities and Structures of Enzymes Bound to Graphene Oxide with a Protein Glue. *Langmuir* **2013,** *29* (50), 15643–15654.

22. Shen, J.; Shi, M.; Yan, B.; Ma, H.; Li, N.; Hu, Y.; Ye, M. Covalent Attaching Protein to Graphene Oxide via Diimide-Activated Amidation. *Colloids Surfaces B: Biointerf.* **2010,** *81* (2), 434–438.

23. Shen, J.; Yan, B.; Shi, M.; Ma, H.; Li, N.; Ye, M. Synthesis of Graphene Oxide-Based Biocomposites Through Diimide-Activated Amidation. *J. Colloid Interface Sci.* **2011,** *356* (2), 543–549.

24. Chen, Y.; Xianyu, Y.; Wu, J.; Yin, B.; Jiang, X. Click Chemistry-Mediated Nanosensors for Biochemical Assays. *Theranostics* **2016,** *6* (7), 969.

25. Chang, H.-N.; Sarkar, S.; Baker, J. R.; Norris, T. B. Fluorophore and Protein Conjugated Diels-Alder Functionalized CVD Graphene Layers. *Opt. Mater. Express* **2016,** *6* (10), 3242–3253.

26. Zhang, G.; Ma, J.; Wang, J.; Li, Y.; Zhang, G.; Zhang, F.; Fan, X. Lipase Immobilized on Graphene Oxide as Reusable Biocatalyst. *Ind. Eng. Chem. Res.* **2014,** *53* (51), 19878–19883.

27. Wang, Z.; Zhou, X.; Zhang, J.; Boey, F.; Zhang, H. Direct Electrochemical Reduction of Single-Layer Graphene Oxide and Subsequent Functionalization With Glucose Oxidase. *J. Phys. Chem. C* **2009,** *113* (32), 14071–14075.

28. Hermanova, S.; Zarevucka, M.; Bousa, D.; Pumera, M.; Sofer, Z. Graphene Oxide Immobilized Enzymes Show High Thermal and Solvent Stability. *Nanoscale* **2015,** *7* (13), 5852–5858.

29. Patila, M.; Kouloumpis, A.; Gournis, D.; Rudolf, P.; Stamatis, H. Laccase-Functionalized Graphene Oxide Assemblies as Efficient Nanobiocatalysts for Oxidation Reactions. *Sensors* **2016,** *16* (3), 287.

30. Emadi, F.; Amini, A.; Gholami, A.; Ghasemi, Y. Functionalized Graphene Oxide with Chitosan for Protein Nanocarriers to Protect against Enzymatic Cleavage and Retain Collagenase Activity. *Sci. Rep.* **2017,** *7*.

31. Yang, T.; Liu, L.-h; Liu, J.-w; Chen, M.-L.; Wang, J.-H. Cyanobacterium Metallothionein Decorated Graphene Oxide Nanosheets for Highly Selective Adsorption of Ultra-Trace Cadmium. *J. Mater. Chem.* **2012,** *22* (41), 21909–21916.

32. Cheng, C.; Li, S.; Nie, S.; Zhao, W.; Yang, H.; Sun, S.; Zhao, C. General and Biomimetic Approach to Biopolymer-Functionalized Graphene Oxide Nanosheet Through Adhesive Dopamine. *Biomacromolecules* **2012,** *13* (12), 4236–4246.

33. Mei, K.-C.; Rubio, N.; Costa, P. M.; Kafa, H.; Abbate, V.; Festy, F.; Bansal, S. S.; Hider, R. C.; Al-Jamal, K. T. Synthesis of Double-Clickable Functionalised Graphene Oxide for Biological Applications. *Chem. Commun.* **2015,** *51* (81), 14981–14984.

34. Georgakilas, V.; Tiwari, J. N.; Kemp, K. C.; Perrnan, J. A.; Bourlinos, A. B.; Kim, K. S.; Zboril, R. Noncovalent Functionalization of Graphene and Graphene Oxide for Energy Materials, Biosensing, Catalytic, and Biomedical Applications. *Chem. Rev.* **2016,** *116* (9), 5464–5519.

35. Kumar, C.; Pattammattel, A. Chapter Eleven-BioGraphene: Direct Exfoliation of Graphite in a Kitchen Blender for Enzymology Applications. *Methods Enzymol.* **2016,** *571,* 225−244.

36. Liu, J.; Fu, S.; Yuan, B.; Li, Y.; Deng, Z. Toward a Universal "Adhesive Nanosheet" for the Assembly of Multiple Nanoparticles Based on a Protein-Induced Reduction/ Decoration of Graphene Oxide. *J. Am. Chem. Soc.* **2010,** *132,* 7279−7281.

37. Tian, J.; Yuan, P.-X.; Shan, D.; Ding, S.-N.; Zhang, G.-Y.; Zhang, X.-J. Biosensing Platform Based on Graphene Oxide via Self-Assembly Induced by Synergic Interactions. *Anal. Biochem.* **2014,** *460,* 16−21.

38. Liu, X.; Han, Z.; Li, F.; Gao, L.; Liang, G.; Cui, H. Highly Chemiluminescent Graphene Oxide Hybrids Bifunctionalized by N-(Aminobutyl)-N-(Ethylisoluminol)/ Horseradish Peroxidase and Sensitive Sensing of Hydrogen Peroxide. *ACS Appl. Mater. Interfaces* **2015,** *7* (33), 18283−18291.

39. Chang, H.; Tang, L.; Wang, Y.; Jiang, J.; Li, J. Graphene Fluorescence Resonance Energy Transfer Aptasensor for the Thrombin Detection. *Anal. Chem.* **2010,** *82* (6), 2341−2346.

40. Wang, W.; Ge, L.; Sun, X.; Hou, T.; Li, F. Graphene-Assisted Label-Free Homogeneous Electrochemical Biosensing Strategy based on Aptamer-Switched Bidirectional DNA Polymerization. *ACS Appl. Mater. Interfaces* **2015,** *7* (51), 28566−28575.

41. Zhang, J.; Zhang, F.; Yang, H.; Huang, X.; Liu, H.; Zhang, J.; Guo, S. Graphene Oxide as a Matrix for Enzyme Immobilization. *Langmuir* **2010,** *26* (9), 6083−6085.

42. Jin, L.; Yang, K.; Yao, K.; Zhang, S.; Tao, H.; Lee, S.-T.; Liu, Z.; Peng, R. Functionalized Graphene Oxide in Enzyme Engineering: A Selective Modulator for Enzyme Activity and Thermostability. *ACS Nano* **2012,** *6* (6), 4864−4875.

43. Mathesh, M.; Luan, B.; Akanbi, T. O.; Weber, J. K.; Liu, J.; Barrow, C. J.; Zhou, R.; Yang, W. Opening Lids: Modulation of Lipase Immobilization by Graphene Oxides. *ACS Catalysis* **2016,** *6* (7), 4760−4768.

44. Sun, C.; Walker, K. L.; Wakefield, D. L.; Dichtel, W. R. Retaining the Activity of Enzymes and Fluorophores Attached to Graphene Oxide. *Chem. Mater.* **2015,** *27* (12), 4499−4504.

45. Huang, C.; Bai, H.; Li, C.; Shi, G. A Graphene Oxide/Hemoglobin Composite Hydrogel for Enzymatic Catalysis in Organic Solvents. *Chem. Commun.* **2011,** *47* (17), 4962−4964.

46. Baptista-Pires, L.; Pérez-López, B.; Mayorga-Martinez, C. C.; Morales-Narváez, E.; Domingo, N.; Esplandiu, M. J.; Alzina, F.; Torres, C. M. S.; Merkoçi, A. Electrocatalytic Tuning of Biosensing Response Through Electrostatic or Hydrophobic Enzyme−Graphene Oxide Interactions. *Biosensors Bioelectron.* **2014,** *61,* 655−662.

47. Liu, C.; Alwarappan, S.; Chen, Z.; Kong, X.; Li, C.-Z. Membraneless Enzymatic Biofuel Cells Based on Graphene Nanosheets. *Biosensors Bioelectron.* **2010,** *25* (7), 1829−1833.

48. Novak, M. J.; Pattammattel, A.; Koshmerl, B.; Puglia, M.; Williams, C.; Kumar, C. V. "Stable-on-the-Table" Enzymes: Engineering the Enzyme−Graphene Oxide Interface for Unprecedented Kinetic Stability of the Biocatalyst. *ACS Catalysis* **2015,** *6* (1), 339−347.

49. Zore, O. V.; Pattammattel, A.; Gnanaguru, S.; Kumar, C. V.; Kasi, R. M. Bienzyme—Polymer—Graphene Oxide Quaternary Hybrid Biocatalysts: Efficient Substrate Channeling Under Chemically and Thermally Denaturing Conditions. *ACS Catalysis* **2015,** *5* (9), 4979—4988.

50. Pattammattel, A.; Williams, C. L.; Pande, P.; Tsui, W. G.; Basu, A. K.; Kumar, C. V. Biological Relevance of Oxidative Debris Present in As-Prepared Graphene Oxide. *RSC Adv.* **2015,** *5* (73), 59364—59372.

51. Deepachitra, R.; Chamundeeswari, M.; Santhosh kumar, B.; Krithiga, G.; Prabu, P.; Pandima Devi, M.; Sastry, T. P. , Osteo Mineralization of Fibrin-Decorated Graphene Oxide. *Carbon* **2013,** *56,* 64—76.

52. Liu, H.; Cheng, J.; Chen, F.; Bai, D.; Shao, C.; Wang, J.; Xi, P.; Zeng, Z. Gelatin Functionalized Graphene Oxide for Mineralization of Hydroxyapatite: Biomimetic and In Vitro Evaluation. *Nanoscale* **2014,** *6* (10), 5315—5322.

53. Kelly, S. M.; Jess, T. J.; Price, N. C. How to Study Proteins by Circular Dichroism. *Biochim. Biophys. Acta (BBA)-Proteins Proteomics* **2005,** *1751* (2), 119—139.

54. Kong, J.; Yu, S. Fourier Transform Infrared Spectroscopic Analysis of Protein Secondary Structures. *Acta Biochim. Biophys. Sinica* **2007,** *39* (8), 549—559.

55. Kim, J.-H.; Yoon, J.-Y. Protein Adsorption on Polymer Particles. *Citeseer* **2002,** *1,* 4373.

56. Shen, H.; Liu, M.; He, H.; Zhang, L.; Huang, J.; Chong, Y.; Dai, J.; Zhang, Z. PEGylated Graphene Oxide-Mediated Protein Delivery for Cell Function Regulation. *J. Appl. Mater. Interfaces* **2012,** *4,* 6317—6323.

57. Swathi, R. S.; Sebastian, K. L. Long Range Resonance Energy Transfer From a Dye Molecule to Graphene Has (distance)[sup -4] Dependence. *J. Chem. Phys.* **2009,** *130* 086101-3.

58. Lakowicz, J. *Principles of Fluorescence Spectroscopy;* Kluwer Academic/Plenum Publishers: New York, Boston, Dordrecht, London, Moscow, **1999.**

59. Romanowska, J.; Kokh, D. B.; Wade, R. C. When the Label Matters: Adsorption of Labeled and Unlabeled Proteins on Charged Surfaces. *Nano Lett.* **2015,** *15* (11), 7508—7513.

60. Park, J. S.; Cho, S. M.; Kim, W.-J.; Park, J.; Yoo, P. J. Fabrication of Graphene Thin Films Based on Layer-by-Layer Self-Assembly of Functionalized Graphene Nanosheets. *J. Appl. Mater. Interfaces* **2011,** *3,* 360—368.

61. Li, D.; Muller, M. B.; Gilje, S.; Kaner, R. B.; Wallace, G. G. Processable Aqueous Dispersions of Graphene Nanosheets. *Nat. Nano* **2008,** *3,* 101—105.

62. Vickery, J. L.; Patil, A. J.; Mann, S. Fabrication of Graphene—Polymer Nanocomposites With Higher-Order Three-Dimensional Architectures. *Adv. Mater.* **2009,** *21,* 2180—2184.

63. Zhu, Y.; Cai, W.; Piner, R. D.; Velamakanni, A.; Ruoff, R. S. Transparent Self-Assembled Films of Reduced Graphene Oxide Platelets. *Appl. Phys. Lett.* **2009,** *95* 103104-3.

64. Pattammattel, A.; Deshapriya, I. K.; Chowdhury, R.; Kumar, C. V. Metal-Enzyme Frameworks: Role of Metal Ions in Promoting Enzyme Self-Assembly on α-Zirconium(IV) Phosphate Nanoplates. *Langmuir* **2013,** *29* (9), 2971—2981.

65. Wright, A. K.; Thompson, M. R. Hydrodynamic Structure of Bovine Serum Albumin Determined by Transient Electric Birefringence. *Biophys. J.* **1975,** *15* (2), 137—141.

66. Kumar, C. V.; Chaudhari, A. Proteins Immobilized at the Galleries of Layered A-Zirconium Phosphate: Structure and Activity Studies. *J. Am. Chem. Soc.* **2000,** *122* (5), 830−837.

67. Deshapriya, I. K.; Kumar, C. V. Nanobio Interfaces: Charge Control of Enzyme/inorganic Interfaces for Advanced Biocatalysis. *Langmuir* **2013,** *29* (46), 14001−14016.

68. Zhang, Y.; Wu, C.; Guo, S.; Zhang, J. Interactions of Graphene and Graphene Oxide with Proteins and Peptides. *Nanotechnol. Rev.* **2013,** *2,* 27.

69. Chou, S. S.; De, M.; Luo, J.; Rotello, V. M.; Huang, J.; Dravid, V. P. Nanoscale Graphene Oxide (nGO) as Artificial Receptors: Implications for Biomolecular Interactions and Sensing. *J. Am. Chem. Soc.* **2012,** *134,* 16725−16733.

70. Katsnelson, M. I. Graphene: Carbon in Two Dimensions. *Mater. Today* **2007,** *10,* 20−27.

71. Diederix, R. E. M.; Ubbink, M.; Canters, G. W. Effect of the Protein Matrix of Cytochrome c in Suppressing the Inherent Peroxidase Activity of Its Heme Prosthetic Group. *ChemBioChem* **2002,** *3,* 110−112.

72. Mu, Q.; Su, G.; Li, L.; Gilbertson, B. O.; Yu, L. H.; Zhang, Q.; Sun, Y.-P.; Yan, B. Size-Dependent Cell Uptake of Protein-Coated Graphene Oxide Nanosheets. *J. Appl. Mater. Interfaces* **2012,** *4,* 2259−2266.

73. Wang, S.; Zhang, Y.; Abidi, N.; Cabrales, L. Wettability and Surface Free Energy of Graphene Films. *Langmuir* **2009,** *25,* 11078−11081.

74. Mudhivarthi, V. K.; Bhambhani, A.; Kumar, C. V. Novel Enzyme/DNA/Inorganic Nanomaterials: A New Generation of Biocatalysts. *Dalton Trans.* **2007,** 5483.

75. Kumar, C.; Chaudhari, A. High Temperature Peroxidase Activities of HRP and Hemoglobin in the Galleries of Layered Zr (IV) Phosphate. *Chem. Commun.* **2002,** *20,* 2382−2383.

76. Yao, K.; Tan, P.; Luo, Y.; Feng, L.; Xu, L.; Liu, Z.; Li, Y.; Peng, R. Graphene Oxide Selectively Enhances Thermostability of Trypsin. *ACS Appl. Mater. Interfaces* **2015,** *7* (22), 12270−12277.

77. Mudhivarthi, V. K.; Cole, K. S.; Novak, M. J.; Kipphut, W.; Deshapriya, I. K.; Zhou, Y.; Kasi, R. M.; Kumar, C. V. Ultra-Stable Hemoglobin−Poly (Acrylic Acid) Conjugates. *J. Mater. Chem.* **2012,** *22* (38), 20423−20433.

78. Sanchez-Ruiz, J. M. Enzyme Kinetic Stability. *Biophys. Chem.* **2010,** *148,* 1−15.

79. Zhang, C.; Chen, S.; Alvarez, P. J. J.; Chen, W. Reduced Graphene Oxide Enhances Horseradish Peroxidase Stability by Serving as Radical Scavenger and Redox Mediator. *Carbon* **2015,** *94,* 531−538.

80. Greenfield, P. F.; Kittrell, J. R.; Laurence, R. L. Inactivation of Immobilized Glucose Oxidase by Hydrogen Peroxide. *Anal. Biochem.* **1975,** *65,* 109−124.

81. Pumera, M. Graphene in Biosensing. *Mater. Today* **2011,** *14* (7−8), 308−315.

82. Lu, C. H.; Yang, H. H.; Zhu, C. L.; Chen, X.; Chen, G. N. A Graphene Platform for Sensing Biomolecules. *Angew. Chem.* **2009,** *121* (26), 4879−4881.

83. He, S.; Song, B.; Li, D.; Zhu, C.; Qi, W.; Wen, Y.; Wang, L.; Song, S.; Fang, H.; Fan, C. A graphene Nanoprobe for Rapid, Sensitive, and Multicolor Fluorescent DNA Analysis. *Adv. Funct. Mater.* **2010,** *20* (3), 453−459.

84. Kong, F.-Y.; Gu, S.-X.; Li, W.-W.; Chen, T.-T.; Xu, Q.; Wang, W. A Paper Disk Equipped with Graphene/Polyaniline/Au Nanoparticles/Glucose Oxidase Biocomposite Modified Screen-Printed Electrode: Toward Whole Blood Glucose Determination. *Biosensors Bioelectron.* **2014,** *56,* 77−82.

85. Vashist, S. K.; Luong, J. H. T. Recent Advances in Electrochemical Biosensing Schemes Using Graphene and Graphene-Based Nanocomposites. *Carbon* **2015,** *84,* 519−550.

86. Zhang, Y.; Shen, J.; Li, H.; Wang, L.; Cao, D.; Feng, X.; Liu, Y.; Ma, Y.; Wang, L. *The Chemical Record* **2016,** *16,* 273−294.

87. Song, Y.; Luo, Y.; Zhu, C.; Li, H.; Du, D.; Lin, Y. Recent Advances in Electrochemical Biosensors Based on Graphene Two-Dimensional Nanomaterials. *Biosensors Bioelectron.* **2016,** *76,* 195−212.

88. Liu, Z.; Robinson, J. T.; Sun, X.; Dai, H. PEGylated Nanographene Oxide for Delivery of Water-Insoluble Cancer Drugs. *J. Am. Chem. Soc.* **2008,** *130* (33), 10876−10877.

89. Depan, D.; Shah, J.; Misra, R. D. K. Controlled Release of Drug From Folate-Decorated and Graphene Mediated Drug Delivery System: Synthesis, Loading Efficiency, and Drug Release Response. *Mater. Sci. Eng. C* **2011,** *31* (7), 1305−1312.

90. Wen, H.; Dong, C.; Dong, H.; Shen, A.; Xia, W.; Cai, X.; Song, Y.; Li, X.; Li, Y.; Shi, D. *Small* **2012,** *8,* 760−769.

91. Kim, H.; Lee, D.; Kim, J.; Kim, T.-i; Kim, W. J. Photothermally Triggered Cytosolic Drug Delivery via Endosome Disruption Using a Functionalized Reduced Graphene Oxide. *ACS Nano* **2013,** *7* (8), 6735−6746.

92. Kotchey, G. P.; Hasan, S. A.; Kapralov, A. A.; Ha, S. H.; Kim, K.; Shvedova, A. A.; Kagan, V. E.; Star, A. A Natural Vanishing Act: The Enzyme-Catalyzed Degradation of Carbon Nanomaterials. *Acc. Chem. Res.* **2012,** *45* (10), 1770−1781.

93. Yang, K.; Zhang, S.; Zhang, G.; Sun, X.; Lee, S.-T.; Liu, Z. Graphene in Mice: Ultrahigh In Vivo Tumor Uptake and Efficient Photothermal Therapy. *Nano Lett.* **2010,** *10* (9), 3318−3323.

94. Tenzer, S.; Docter, D.; Kuharev, J.; Musyanovych, A.; Fetz, V.; Hecht, R.; Schlenk, F.; Fischer, D.; Kiouptsi, K.; Reinhardt, C., et al. Rapid Formation of Plasma Protein Corona Critically Affects Nanoparticle Pathophysiology. *Nat. Nano* **2013,** *8,* 772−781.

95. Salvati, A.; Pitek, A. S.; Monopoli, M. P.; Prapainop, K.; Bombelli, F. B.; Hristov, D. R.; Kelly, P. M.; Aberg, C.; Mahon, E.; Dawson, K. A. Transferrin-Functionalized Nanoparticles Lose Their Targeting Capabilities When a Biomolecule Corona Adsorbs on the Surface. *Nat. Nano* **2013,** *8,* 137−143.

96. Wilhelm, S.; Tavares, A. J.; Dai, Q.; Ohta, S.; Audet, J.; Dvorak, H. F.; Chan, W. C. W. Analysis of Nanoparticle Delivery to Tumours. *Nat. Rev. Mater.* **2016,** *1,* 16014.

97. Pattammattel, A.; Kumar, C. V. Kitchen Chemistry 101: Multigram Production of High Quality Biographene in a Blender with Edible Proteins. *Adv. Funct. Mater.* **2015,** *25* (45), 7088−7098.

Nanotoxicity of graphene

CHAPTER OUTLINE

9.1 Summary 187
9.2 Introduction 188
9.3 Challenges in Experimental Design of Nanotoxicity Assays 189
9.4 Methods for Nanotoxicity Assessments 192
9.5 In Vitro Toxicity of Graphene..192
9.6 In Vivo Toxicity of Graphene 198
9.7 Conclusions..198
References 202

9.1 SUMMARY

Nanosafety has been a major concern in the 21st century because of the fascinating properties of nanoscale materials and their impact on the environment. The potential risk of using nanomaterials in real life is a big concern among scientists and environmentalists but the general public is not fully appraised of the potential safety or dangers of these materials. The toxicity of nanomaterials has been assessed in cell and animal models, using protocols from molecular toxicology but the methods and implementations may not be adequate or even relevant in some instances. The assessments are done to evaluate the cytotoxic and genotoxic impacts of these materials by following the levels of specific biomarkers. Despite strong efforts to determine the toxicity of nanomaterials, there still is not a predictive tool available to assess the risk of nanomaterials and its relation to the chemical composition of the material. This chapter, therefore, begins with background information on nanotoxicity measures, their strengths and drawbacks, and certain recommendations that might be helpful for more critical evaluation of the data. Next, an overview of biological assays used to evaluate various levels of nanoparticle toxicity, from cell viability to mutations, will be discussed. These particular sections are designed for use by material chemists who are not too familiar with biological assay protocols. The discussion and overview on graphene and graphene oxide (GO) toxicity in cell and animal models will then follow the above topics, and two opposite opinions among graphene researchers will be presented. Our own data on graphene produced using animal sera revealed low toxicity. Graphene toxicity studies are expected to continue into the future because of its

Introduction to Graphene. DOI: http://dx.doi.org/10.1016/B978-0-12-813182-4.00009-X

mass production options, low cost, lower environmental impact, and greater feasibility to make devices. Careful examination of the graphene toxicity at realistic dosage levels, under a variety of conditions that are relevant to real-life situations, is vital for designing safety standards for the common use of this wonder material.

9.2 INTRODUCTION

One of the major concerns in using nanomaterials for real-life applications is their potential toxicity.[1] There is currently a pressing need to assess the cytotoxicity, environmental toxicity, and genotoxicity of nanomaterials due to their rise in production in the last decade.[2] Even after 3−4 decades of research in nanomaterials, little is known about their adverse outcomes and currently the concept of nano environmental health and safety (nano-EHS) is being introduced.[3,4] Most of these studies have centered on nanoparticles of titanium dioxide (TiO_2) and zinc oxide (ZnO_2) because of their social and economic impacts resulting from their widespread use in cosmetics and sunscreen creams.[5,6] Toxicology studies of black carbon particles have also been investigated in detail, because they are used in coating and as a filler material in tires.[7] Engineered nanomaterials can have many different synthetic routes, sizes, shapes, compositions, impurities, and several other features, making it nearly impossible to generalize the adverse effects of nanomaterials.[8,9]

Various mechanisms for the toxicity of nanoparticles have been proposed in the literature (Fig. 9.1).[10] Nanomaterials induce the generation of reactive oxygen (ROS) or reactive nitrogen species (RNS) in cells causing oxidative stress.[11] These reactive species can also react with DNA, causing oxidation to nucleobases, in particular guanine, forming mutagenic changes in DNA such as the production of 8-oxoguanosine.[12] Nanoparticles can also stimulate an inflammatory response in immune cells, and induce cytokine production.[13] Cytokines are the set of proteins generated from immune cells which are capable of triggering chronic inflammation and disease conditions such as rheumatoid arthritis.[13] However, the long-term effects of nanomaterial toxicity are not yet well understood.

Graphene derivatives have also shown cytotoxic, genotoxic and inflammatory responses in biological models.[14] Like other studies at nano−bio interfaces, nanotoxicity research is also dominated by GO-based models.[15] In this chapter, we attempt to focus on unoxidized graphene-based assays, but selective investigations of GO are included. In vitro studies using macrophages, lung, kidney, liver, epithelial, and spleen cells are reviewed, which have a greater chance of interacting with graphene derivatives by inhalation or physical contact. Layered materials like graphene can have different sizes, number of layers, and oxidation levels, which are important in deciding the cellular responses.[15] Before discussing the

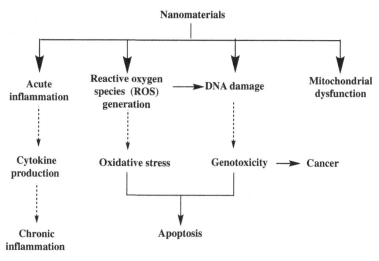

FIGURE 9.1

Scheme showing the major primary and secondary effects of nanomaterials in biological systems.

details of these studies, an introduction to the challenges and experimental methods related to nanotoxicity assays is given.

9.3 CHALLENGES IN EXPERIMENTAL DESIGN OF NANOTOXICITY ASSAYS

Early studies on nano- or micron-sized materials were originally studied in air pollution models, where the potential risks associated with asbestos and silicates were revealed.[16] Besides, in vivo and in vitro toxicity models of commercially used nanomaterials (such as carbon black, TiO_2) were subjects of interest.[5] Although the carcinogenicity of asbestos has been established in the past, disagreement in the findings for other engineered nano materials pose challenges in this field. These results often lead to disagreements between academic researchers and the nanoparticle industry.[17–20] Similarly, toxicity studies on carbon nanotubes, nanofibers, and fullerenes also lack agreement in several in vitro and in vivo models.[7] A few reasons are noted for these discrepancies in various nanotoxicity assays.[21]

Poor understanding of the bio—nano interface, its complexity as well as heterogeneity are some of the primary reasons for poor consistency among different nanotoxicity assays. Unfavorable interactions between the nanoparticles and biological fluids, which are rich in serum proteins, cause changes in size, morphology, and colloidal properties.[22] This common problem is referred to as protein

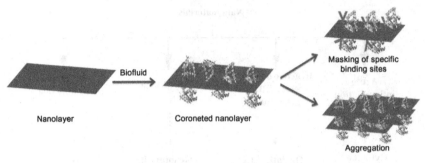

FIGURE 9.2

Biocorona formation around 2-D layers and its consequences.

corona formation (Fig. 9.2).[23] Due to these events, the stability of the nanoparticle dispersions in biofluids is limited.[24] Studies suggest that protein corona masks the toxicity of nanomaterials,[25,26] while in other cases it enhances the toxicity responses and intake of nanoparticles.[27,28] The extent of aggregation of nanoparticles is also dependent on the type and percentage of serum used in the studies, which leads to variable results in toxicity assays.[29,30] Also, uncontrollable interactions of nanoparticles and proteins lead to the precipitation of the particles and the portion of material interacting with the cells or tissue is lower than the dose. This leads to the overestimation of dosage range in assays. Another experimental error in these studies comes from contamination or impurities present in the material.

Impurities in nanomaterials can be present in the dispersion as a result of synthesis conditions. For example, GO synthesis via Hummer's method forms highly oxidized polyaromatic hydrocarbons, called oxidative debris, as well as manganese oxide nanoparticles, which can have biological reactions.[31,32] The metallic impurities are especially hyperactive in biological pathways in comparison to the pure carbon-based layers. Another possible problem is the contamination of nanoparticles by microorganisms or endotoxins from ambient laboratory conditions.[33] General sterilizing steps such as autoclaving, ethanol wash, or UV exposure might affect the properties of the particles and thus, nanoparticle purification from these is crucial.

Finally, the presence of nanoparticles could possibly interfere with bioassays during spectroscopic measurements.[34] Optical scattering and absorption from the nanomaterials causes baseline shifts or peak shifts in absorbance as well as fluorescence measurements.[34] This would result in false-positive readings during absorbance measurements because absorbance/scattering by the nanoparticles is added to the baseline shift in the spectral reading. Most of the biological assays monitor increases in dye absorbance during the assay. Furthermore, different doses of nanoparticles will have different extents of baseline shift, which can again induce errors in readouts from the same set. Assays based on fluorescence measurements would also have such spectroscopic errors in addition to baseline shifts. Nanomaterials, especially carbon-based systems, are strong fluorescent quenchers, which could cause fluorescent emission from the reporter molecule to be underestimated. Conversely, systems such as quantum dots or carbon dots are strongly fluorescent nanosystems, which can also contribute to the fluorescence readouts. A few recommendations to reduce these errors in biological assays from our experience and literature are listed below.

1. Complete characterization of the material under cell culture media or in biofluids is needed to account for protein aggregation. Careful monitoring of the size and stability of the particles over time, with respect to the period of cell/animal exposure, is essential. Dynamic light scattering and zeta potential analysis are suitable for nanoparticle characterization in protein-rich media.[35]

2. Perform control experiments of biological assays with and without serum in cell culture media. This study accounts for any role of serum proteins in bioassays. Serum proteins may be necessary for efficient cell growth. In those cases, the media can be switched to serum-free conditions after sufficient confluence, prior to the exposure of nanoparticles in the assay.

3. Appropriate sterilization of the nanoparticles by a suitable method is essential. There is no universal method for sterilization of nanoparticles due to the variety of conditions the particles cannot tolerate. Methods that would not affect the integrity of the nanoparticles have to be identified and implemented for each system. Synthesis and purification of graphene under completely sterile conditions were adopted in our laboratory when common sterilization methods were shown to be affecting the integrity of the samples.[36]

4. Complete characterization of the chemical composition of the system is essential. Studies find that trace amounts of metallic impurities can be responsible for various biological and nonbiological reactions. Thus, characterization for trace metal analysis such as inductively coupled plasma—mass spectrometry/optical emission spectroscopy (ICP-MS or ICP-OES) or elemental mapping of statistically significant numbers of particles using electron microscopy may be used. The presence of any endotoxins in the material also should be accounted for as they can cause toxic effects.[33]

5. Baseline correction in spectroscopic measurements during bioassays is another important factor that should be taken into consideration.[37] For absorption-based systems, blanks with nanoparticles alone can be used for correction of any contribution from scattering or absorption. Nanoparticles mixed with dead cells can also be used to account for any changes in interference due to biophysical interactions. In fluorescence assays, measurement of the label alone in the presence of nanoparticles can be used to account for quenching or inner filter effect. Performing bioassays after the removal of existing culture media containing nanoparticles could result in less interference in spectroscopy measurements. Another possibility is the use of nonspectroscopic quantifications such as western blotting, although they are time consuming.

9.4 METHODS FOR NANOTOXICITY ASSESSMENTS

The cell or animal responses to nanoparticle exposure are quantitated using various bioassays including, but not limited to, cell viability, metabolic rate, ROS levels, cytokine levels, apoptosis, and/or gene regulation assays.[9,38] Classic chemical toxicology techniques are utilized for assessment of potential toxicity of nanomaterials. In vitro experiments offer rapid assessment of acute toxicity from nanomaterials. These can easily be executed for various cellular events of interest. Table 9.1 describes the commonly used toxicity bioassays in nanotechnology.

9.5 IN VITRO TOXICITY OF GRAPHENE

The toxicity of graphene, and its analogues, to several cell lines is examined using a combination of techniques described above (Table 9.1). The challenge in analyzing the effect of graphene, rather than GO, is the poor dispersive nature of the unoxidized layers in aqueous conditions. Dispersion of graphene synthesized by chemical vapor deposition or mechanical exfoliation in biocompatible surfactants, or polymers, is used in cell studies (Table 9.2). These results show that graphene causes ROS generation inside the cells and eventually leads to apoptosis at around 100 μg/mL dose levels.[69] Interestingly, these responses are also recorded for GO, and reduced GO, where mitochondrial dysfunction was found to be induced by the particles. Size-dependent genotoxicity of GO in human lung cancer cells (A549), was evaluated using comet assay.[70] The mutagenic behavior of GO (10–100 μg/mL and 4 mg/kg) during DNA replication was demonstrated in in vivo and in vitro models and the mutagenicity was demonstrated to be equivalent to classic small-molecule carcinogens.[71] On the contrary, in vitro studies of GO that are free of metallic impurities and endotoxins show no observable cytotoxicity or genotoxicity to mouse lung epithelial cells, even at higher doses (5–200 μg/mL). However, all of these studies agree that smaller GO sheets have higher cytotoxic and genotoxic effects. Unfortunately, the toxicity of unoxidized graphene is not compared in these reports.

Graphene produced by liquid-phase exfoliation in animal sera displayed no observable cytotoxicity to cells or *Caenorhabditis elegans*, even after 10–14 days of exposure at doses below 100 μg/mL (Fig. 9.3).[36] A reduction in intracellular metabolism after more than 10 days of exposure for doses above 100 μg/mL suggested the dose-dependent toxicity of unoxidized graphene. However, an isolated fraction of smaller graphene sheets (below 200 nm) caused cytotoxic responses by 10–60 μg/mL after exposure for over a week. Moreover, no toxicity from graphene was observed for *C. elegans* at any concentrations. This study implies that the unoxidized graphene derivatives are better than the oxidized derivatives in a biological context. The origin of toxicity of those derivatives, however, is a subject of controversy. Selected examples of cytotoxic and genotoxic effects of graphene and GO are presented in Table 9.2.

Table 9.1 Commonly Used Bioassays to Quantitate the Potential Toxicity of Nanomaterials

Bioassay	Marker of	Background	Measurement Technique
Lactate dehydrogenase (LDH) assay[39]	• Cell membrane integrity • Lytic cell death	Up on lytic cell death LDH releases to extracellular matrix, whose activity can be monitored its activity to NADH	Absorption or fluorescence spectroscopy
Microculture tetrazolium assays (MTA)[40]	• Cell metabolic activity • Mitochondrial respiration	Mitochondrial succinic dehydrogenases activity is monitored using bioreduction of tetrazolium salts such as 3-(4,5-dimethylthiazol-2-yl)-2,5-diphenyltetrazolium bromide (MTT) or 3-(4,5-dimethylthiazol-2-yl)-5-(3-carboxymethoxyphenyl)-2-(4-sulfophenyl)-2H-tetrazolium (MTS)	Absorption spectroscopy
Trypan blue exclusion assay[41]	Cell death	• Dead cells exclusively take up trypan blue (an azo dye) • Dead cells/live cells ratio is counted after staining	Light microscope
Neutral red assay[42]	Lysosomal activity	Selective uptake of basic red 5 and toluylene red by live cells and their subsequent incorporation into lysosomes	Absorption spectroscopy (540 nm)
DNA laddering[43]	Apoptosis	Fragmentation of DNA by caspase-activated Dnase during apoptosis is monitored	Gel electrophoresis
Caspase-3 activity[44]	Apoptosis	• Capsase-3 is a protease family enzyme which is an executioner of cell apoptosis • Fluorogenic polypeptide chains are used to monitor caspase activity after material treatment	Fluorescence spectroscopy
TdT dUTP nick end labeling (TUNEL) and apostain assays[45]	Apoptosis	• DNA fragmentation during apoptosis was monitored using labeled DNA oligomers • DNA fragments due to apoptosis were selectively identified by thermal denaturation, followed by apostain antibody	• Fluorescence spectroscopy • Light microscopy • Antibody assays

(Continued)

Table 9.1 Commonly Used Bioassays to Quantitate the Potential Toxicity of Nanomaterials *Continued*

Bioassay	Marker of	Background	Measurement Technique
DCF (dichlorofluorescein) assay[46]	• ROS levels • Oxidative stress	Reactive oxygens oxidize nonfluorescent 2,7–dichlorofluorescin diacetate to florescent DCF inside the cells	Fluorescence spectroscopy
Glutathione/oxidized glutathione (GSH/GSSG) assay[47]	• ROS levels • Oxidative stress	• Abnormal levels of ROS shifts to GSH/GSSG equilibrium towards GSSG • Dyes with enhanced emission at oxidative conditions are used for ratio-metric fluorescence sensing	• Fluorescence spectroscopy • Confocal imaging • Flow cytometry
Superoxide dismutase (SOD) assay[48]	Oxidative stress	• SOD is responsible for the clearance of superoxide species in the cells • SOD inhibition of superoxide-induced reduction reactions are monitored	Absorption or fluorescence spectroscopy
Malondialdehyde (MDA) assay[49]	• Lipid peroxidation • Oxidative stress	• MDA is the end product of lipid peroxidation • MDA is quantified using thiobarbituric acid (TBA) which forms products that are colored or fluorescent	Absorption or fluorescence spectroscopy
Caspase-1 assay[50]	Inflammation	• Caspase is the protease responsible for the production of proinflammatory cytokine, interleukin 1 beta (IL-1-β) • Activity of the enzyme is measured using a fluorogenic polypeptide	Florescence spectroscopy
Clonogenic assay or colony-forming efficiency (CFE)[51]	Cell proliferation	Following the efficiency of cell division in the presence and absence of the material	Light microscope
DNA content assay[52]	Cell proliferation	Cell cycle during mitosis was arrested at metaphase and the amount of DNA present is compared	Electrophoresis and various staining DNA techniques

Assay	Application	Principle	Detection method
[³H] thymidine assay[53]	Cell proliferation/ DNA quantification	Radiolabeled (^{3}H) thymidine incorporates to the chromosomes during the mitosis step	Scintillation counter
Bromodeoxyuridine (BrdU) assay[54]	Cell proliferation/ DNA quantification	BrdU associates with DNA during cell division and it is detected by antibody assays	Fluorescence spectroscopy
Measurement of oxidized guanine bases[55]	DNA damage	High ROS generation oxidizes neighboring DNA and amine groups in guanine are oxidized	High-performance liquid chromatography (HPLC) and quantitative PCR (polymerase chain reaction)
Micronucleus test[56]	DNA mutation/ genotoxicity	Genotoxic substances induce formation of fragmented chromosomal nuclei, called micronucleus in the cytoplasm, which can be visualized under a light microscope	Light microscopy
Chromosomal aberration test[57]	Genotoxicity	Chromosomal defects such as gaps, breaks, exchanges on exposure to toxins such as recorded from microscopy images	Light microscopy
Comet assay (single-cell gel electrophoresis, SCGE)[58]	DNA mutation/ genotoxicity	Damaged DNA fragments move faster than the intact DNA during electrophoresis	Light microscopy and image-processing softwares
Northern blot analysis, ribonuclease protection assays (RPA), quantitative real-time polymerase chain reaction (qRT-PCR), PCR arrays, and microarrays[59]	Gene upregulation or downregulation	Any changes in gene regulation during cell cycling in the presence or absence of materials are compared	Fluorescence or absorption detectors

Table 9.2 Selected In Vitro Toxicity Results of Graphene and Graphene Oxide Derivatives

Material	Properties	Model	Dose	Results
Graphene (CVD) [60]	3–5 layers Charge = −36 mV (in water)	Neuronal cells (PC12)	0.1–100 µg/mL	ROS generation, apoptosis induction
Graphene exfoliated in animal sera	3–5 layers Stable for months in cell culture media	Human kidney cells (HEK293T) and human lung cancer cells (H1299)	50–300 µg/mL	• Low metabolic activity of the cells only at concentrations above 200 µg/mL, after 10 days of exposure • Graphene sheets below 200 nm in size in the dispersion showed high cytotoxicity to the cells • No observable effects on nematode growth or reproduction
Graphene from Kinik Company (Taiwan), dispersed in 1% Pluronic F108 [61]	100–200 nm size, 3–4 nm thickness	Human breast cancer cells	10–80 µg/mL	No influence on cell viability Inhibition of migration and invasion of the cells
Pristine graphene in 1% F108 Pluronic [62]	2–3 nm thickness	RAW 264.7 macrophages	0–100 µg/mL	Pristine graphene can induce cytotoxicity through the depletion of mitochondrial membrane potential Increase of ROS leading to the activation of MAPK and TGF-β Activation of caspase-3 and PARP proteins resulting in apoptosis
Graphene by arc discharge evaporation of graphite [63]	∼ 0.4 nm thickness	Murine macrophage RAW 264.7 cells	50 µg/mL	ROS-mediated apoptosis High levels of cytokine production of pristine graphene compared to oxidized graphene (using 2 M HNO_3)

Material	Properties	Cell/organism	Dose	Effect
Graphene from Kinik Company (Taiwan), dispersed in 1% Pluronic F108[64]	100–200 nm size, 3–4 nm thickness, Zeta potential = −21 mV	Murine macrophage RAW 264.7 cells	20 µg/mL (subcytotoxic dose)	Increase in cytokine production activated by NFκB pathway. Decrease in phagocytosis ability of the macrophages
Graphene platelets from cheaptubes.com[65]	1–10 layers, 1–10 µm, surface area ~100 m²/g	Human monocytic cell line (THP-1)	5 µg/cm²	Increased proinflammatory cytokine production
Fluorinated graphene (F-G)[66]	Three different formulations with varying F content—1.5%, 42.6%, 50.7%	Lung cancer cells (A549)	0–400 µg/mL	Dose-dependent cytotoxicity of fluorinated graphene with greater cytotoxicity for graphene containing higher mono-fluoro substituted carbon atoms
rGO[67]	Size: b5 µm	Aspergillus niger, Fusarium oxysporum, Aspergillus oryzae	0–500 µg/mL	IC50 value is between 50 and 100 µg/mL. rGO shows good antifungal activity
Grade C graphene from XGSciences[68]	Particle diameter: 2 µm	Proteobacteria and Bacteroidetes	300 mg/L	Significant reduction of the microbial community metabolic activity

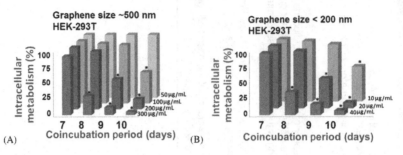

FIGURE 9.3

(A) Relative metabolic activity of human embryonic kidney cells (HEK293T) exposed to several doses of graphene produced in 10% bovine serum using a kitchen blender.
(B) Metabolic activity of HEK cells exposed to smaller graphene sheets shows toxicity at lower doses than as prepared graphene ($*p < 0.05$).[36]

9.6 IN VIVO TOXICITY OF GRAPHENE

Mouse and nematode models are often used to assess toxicity in support of cellular studies of graphene derivatives. Studies suggest that both oxidized and unoxidized graphene have oxidative and inflammatory effects in mice and *C. elegans*. Unfortunately, other studies report alternative results that toxic effects of graphene in animals is minimal at lower doses (Table 9.3). As discussed in the case of in vitro toxicity assessments, graphene toxicity to laboratory animals is also inconclusive. Besides, fewer studies focus on the toxicity of unoxidized graphene than GO and the dispersive nature of graphene is not thoroughly examined. Stable graphene dispersions produced in bovine serum or sodium cholate in our lab exhibited negligible cytotoxicity and reproducible toxicity to *C. elegans*, below $100 \, \mu g/mL$ (Fig. 9.4).[36] At the maximum exposure level ($500 \, \mu g/mL$) a 40% reduction in the survival rate of the nematodes was observed after more than 10 days of exposure. Interestingly, a change in exfoliation medium from bovine serum to sodium cholate did not make any difference to the toxicity of graphene. These findings suggest that the processing of graphene may have a role in toxicology studies. Notably, some responses from mouse lungs are due to the blockage of the airways, rather than the toxicity of the materials.[74] More in vivo studies are required to establish any toxicity of graphene, preferably by using high-purity, stable graphene dispersions. Current findings in this area are listed in Table 9.3.

9.7 CONCLUSIONS

Nanosafety remains a growing topic as more and more unique features of nanosystems are unraveled. The implementation of nanomaterials in everyday life, including graphene, will require nanomaterials to pass critical toxicity

Table 9.3 In Vivo Toxicity Assessments of Graphene Derivatives

Material	Properties	Model	Dose	Results
Graphene platelets (sonication in 2% Pluronic F108)[72]	Thickness = 5 nm (5–15 layers)	Mice	50 µg/animal	No direct toxicity from graphene Well dispersed graphene was less toxic than aggregates
Graphene nanoplatelets (plasma exfoliated, dispersed in 0.5% BSA)[65]	10 nm thick 5 µm in size	Mice	50 µg/animal (aspiration) 5 µg/animal (injection)	Lung and pleural inflammation Frustrated phagocytosis ROS, proinflammatory cytokine activation
Graphene platelets from cheaptubes.com[65]	1–10 layers 1–10 µm surface area ~100 m²/g	Mice	50 µg per mouse	High proinflammatory cytokine production Frustrated phagocytosis Higher inflammation than equal dose of carbon black suggests the later size dependence
Graphene nanoplatelets from cheaptubes.com[73]	1–4 layers 2 µm size Zeta potential = −13 mV	Mice	0.3–1 mg/mouse for acute response and 3 mg/mouse for chronic response	High acute inflammation and LDH levels Translocation of graphene from lungs to mediastinal lymph nodes after 1–4 weeks No chronic immunogenicity of graphene
Aggregated graphene (flocculation of graphite)[74]	3–5 nm thickness when dispersed in 2% pluronic acid	Mice	50 µg/animal	Lodges the airways in lung and induces local fibrotic response
Pristine graphene purchased from Skyspring Nanomaterials, Houston, TX, USA[75]	1–5 nm thickness 4 µm size	Chicken embryo model	50–10,000 µg/L	No effect of graphene in body weight, organ weight, or serum composition Downregulation of DNA synthesis in brain
NanoGraphene Sheets functionalized with polyethylene glycol (NGS-PEG)[76]	Size: 10–50 nm; single or bilayered sheets	Tumor bearing Balb/c mice	Intravenous administration Single dose of 20 mg/kg	High tumor build up, no sign of abnormalities on the kidney, spleen, heart, liver, and lung. Gradual elimination. Low uptake by RES
rGO (Hummers/ thermal reduction)[77]	Bilayer 0.2–5 µm	Mice	250 µg/kg	Less platelet aggregation compared to GO

(Continued)

Table 9.3 In Vivo Toxicity Assessments of Graphene Derivatives *Continued*

Material	Properties	Model	Dose	Results
GO and rGO, Hummers method and hydrazine solution used for reduction[78]	0.8 nm thick sheets	Balb/C mice	20 mg/kg	Toxicity caused due to changes in gene profile. Kidney and liver cells damaged. Apoptotic molecular markers upregulated
GO Hummers method[79]	1.0 nm thickness	Caenorhabditis elegans	1 mg/L	Exposure to GO caused damage to organs due to the deregulation of proteins and receptors in the Wnt/ β-catenin signaling pathway
Multilayer grapheme (sonication of thermally expanded graphite)[80]	30 layers	C. elegans	50–250 μg/mL	No effect on either longevity or reproduction
GO and manganese-contaminated GO nanoparticles, modified Hummers method[81]	—	Acheta domesticus House cricket	200 mg/kg of food	Many histological changes were found in gut and testis of A. domesticus

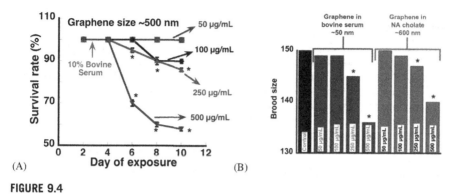

FIGURE 9.4

(A) Survival rate of nematodes (*C. elegans*) exposed to graphene produced in 10% bovine serum using a kitchen blender. (B) Changes in brood size in the nematode colonies exposed to different doses of graphene produced in 10% bovine serum and in 1% sodium cholate (*$p < 0.05$).

assessments. This is important to predict the workplace hazards and environmental effects of the materials. Both in vivo and in vitro models are implemented to examine the toxic effects of nanomaterials. Nanomaterials are often highly heterogeneous in size distribution, shape, and surface chemistry. Thus, predicting or controlling the toxicity of these materials is extremely challenging. Experimental errors due to the poor dispersive nature, serum protein association, presence of impurities, and optical scattering/absorption significantly affects the output of toxicity studies. These can lead to inconsistencies among reports examining similar materials. Graphene is one of the most studied carbon allotropes for potential toxicity in cells and animal models.

The potential toxicity of graphene derivatives is still a matter of controversy. Most of the studies performed using GO show health risks of these materials via an oxidative stress mechanism at fairly high doses (100 μg/mL). Determination of a realistic test dose for graphene in toxicity models needs to be identified. When graphene is used in electronic devices, an extremely low amount of material is necessary. In those contexts, ng/mL concentrations may be a realistic dose. However, if graphene is used in composites or coatings, test doses in the micromolar range would be more suitable. This range would also satisfy workplace hazard levels. Due to the increasing interest in liquid-phase exfoliated graphene, rather than GO, futuristic studies focused on unoxidized graphene are greatly anticipated. For example, graphene produced in animal sera or sodium cholate, with no observable toxicity to nematodes (>200 μg/mL), is a highly promising candidate for applications when compared to the higher toxicity of GO to nematodes.[36] Since the future of implementation of graphene materials in everyday life depends on its toxicity factor, careful studies with reproducible results can be expected in the near future.

REFERENCES

1. Sharifi, S.; Behzadi, S.; Laurent, S.; Laird Forrest, M.; Stroeve, P.; Mahmoudi, M. Toxicity of Nanomaterials. *Chem. Soc. Rev.* **2012,** *41* (6), 2323–2343.
2. Ray, P. C.; Yu, H.; Fu, P. P. Toxicity and Environmental Risks of Nanomaterials: Challenges and Future Needs. *J. Environ. Sci. Health, Part C* **2009,** *27* (1), 1–35.
3. Nel, A.; Zhao, Y.; Mädler, L. Environmental Health and Safety Considerations for Nanotechnology. *Acc. Chem. Res.* **2013,** *46,* 605–606, ACS Publications.
4. Xia, T.; Malasarn, D.; Lin, S.; Ji, Z.; Zhang, H.; Miller, R. J.; Keller, A. A.; Nisbet, R. M.; Harthorn, B. H.; Godwin, H. A. Implementation of a Multidisciplinary Approach to Solve Complex Nano EHS Problems by the UC Center for the Environmental Implications of Nanotechnology. *Small* **2013,** *9* (9–10), 1428–1443.
5. Schilling, K.; Bradford, B.; Castelli, D.; Dufour, E.; Nash, J. F.; Pape, W.; Schulte, S.; Tooley, I.; van den Bosch, J.; Schellauf, F. Human Safety Review of "Nano" Titanium Dioxide and Zinc Oxide. *Photochem. Photobiol. Sci.* **2010,** *9* (4), 495–509.
6. Newman, M. D.; Stotland, M.; Ellis, J. I. The Safety of Nanosized Particles in Titanium Dioxide–and Zinc Oxide–Based Sunscreens. *J. Am. Acad. Dermatol.* **2009,** *61* (4), 685–692.
7. Magrez, A.; Kasas, S.; Salicio, V.; Pasquier, N.; Seo, J. W.; Celio, M.; Catsicas, S.; Schwaller, B.; Forró, L. Cellular Toxicity of Carbon-Based Nanomaterials. *Nano Lett.* **2006,** *6* (6), 1121–1125.
8. Maurer-Jones, M. A.; Gunsolus, I. L.; Murphy, C. J.; Haynes, C. L. Toxicity of Engineered Nanoparticles in the Environment. *Anal. Chem.* **2013,** *85* (6), 3036.
9. Dhawan, A.; Sharma, V. Toxicity Assessment of Nanomaterials: Methods and Challenges. *Anal. Bioanal. Chem.* **2010,** *398* (2), 589–605.
10. Fard, J. K.; Jafari, S.; Eghbal, M. A. A Review of Molecular Mechanisms Involved in Toxicity of Nanoparticles. *Adv. Pharm. Bull.* **2015,** *5* (4), 447.
11. Fu, P. P.; Xia, Q.; Hwang, H.-M.; Ray, P. C.; Yu, H. Mechanisms of Nanotoxicity: Generation of Reactive Oxygen Species. *J. Food Drug Anal.* **2014,** *22* (1), 64–75.
12. Grollman, A. P.; Moriya, M. Mutagenesis by 8-Oxoguanine: An Enemy Within. *Trends Genet.* **1993,** *9* (7), 246–249.
13. Stevenson, R.; Hueber, A. J.; Hutton, A.; McInnes, I. B.; Graham, D. Nanoparticles and Inflammation. *ScientificWorldJournal* **2011,** *11,* 1300–1312.
14. Guo, X.; Mei, N. Assessment of the Toxic Potential of Graphene Family Nanomaterials. *J. Food Drug Anal.* **2014,** *22* (1), 105–115.
15. Seabra, A. B.; Paula, A. J.; de Lima, R.; Alves, O. L.; Duran, N. Nanotoxicity of Graphene and Graphene Oxide. *Chem. Res. Toxicol.* **2014,** *27* (2), 159–168.
16. Stayner, L. T.; Dankovic, D.; Lemen, R. A. Occupational Exposure to Chrysotile Asbestos and Cancer Risk: A Review of the Amphibole Hypothesis. *Am. J. Public. Health.* **1996,** *86* (2), 179–186.
17. Morfeld, P.; Bruch, J.; Levy, L.; Ngiewih, Y.; Chaudhuri, I.; Muranko, H. J.; Myerson, R.; McCunney, R. J. Translational Toxicology in Setting Occupational Exposure Limits for Dusts and Hazard Classification – A Critical Evaluation of a Recent Approach to Translate Dust Overload Findings From Rats to Humans. *Part. Fibre. Toxicol.* **2015,** *12* (1), 3.

18. Levy, L.; Chaudhuri, I. S.; Krueger, N.; McCunney, R. J. Does Carbon Black Disaggregate in Lung Fluid? A Critical Assessment. *Chem. Res. Toxicol.* **2012,** *25* (10), 2001–2006.

19. Morfeld, P.; McCunney, R. J.; Levy, L.; Chaudhuri, I. S. Inappropriate Exposure Data and Misleading Calculations Invalidate the Estimates of Health Risk for Airborne Titanium Dioxide and Carbon Black Nanoparticle Exposures in the Workplace. *Environ. Sci. Poll. Res.* **2012,** *19* (4), 1326–1327.

20. Levy, L. S.; Chaudhuri, I.; Morfeld, P.; McCunney, R. Comments on Induction of Inflammasome-Dependent Pyroptosis by Carbon Black Nanoparticles. *J. Biol. Chem.* **2011,** *286* (38), le17-le17.

21. Kong, B.; Seog, J. H.; Graham, L. M.; Lee, S. B. Experimental Considerations on the Cytotoxicity of Nanoparticles. *Nanomedicine* **2011,** *6* (5), 929–941.

22. Lundqvist, M.; Stigler, J.; Elia, G.; Lynch, I.; Cedervall, T.; Dawson, K. A. Nanoparticle Size and Surface Properties Determine the Protein Corona with Possible Implications for Biological Impacts. *Proc. Natl. Acad. Sci.* **2008,** *105* (38), 14265–14270.

23. Tenzer, S.; Docter, D.; Kuharev, J.; Musyanovych, A.; Fetz, V.; Hecht, R.; Schlenk, F.; Fischer, D.; Kiouptsi, K.; Reinhardt, C. Rapid Formation of Plasma Protein Corona Critically Affects Nanoparticle Pathophysiology. *Nat. Nanotechnol.* **2013,** *8* (10), 772–781.

24. Gebauer, J. S.; Malissek, M.; Simon, S.; Knauer, S. K.; Maskos, M.; Stauber, R. H.; Peukert, W.; Treuel, L. Impact of the Nanoparticle–Protein Corona on Colloidal Stability and Protein Structure. *Langmuir* **2012,** *28* (25), 9673–9679.

25. Docter, D.; Bantz, C.; Westmeier, D.; Galla, H. J.; Wang, Q.; Kirkpatrick, J. C.; Nielsen, P.; Maskos, M.; Stauber, R. H. The Protein Corona Protects Against Size- and Dose-Dependent Toxicity of Amorphous Silica Nanoparticles. *Beilstein J. Nanotechnol.* **2014,** *5,* 1380–1392.

26. Hu, W.; Peng, C.; Lv, M.; Li, X.; Zhang, Y.; Chen, N.; Fan, C.; Huang, Q. Protein Corona-Mediated Mitigation of Cytotoxicity of Graphene Oxide. *ACS Nano* **2011,** *5* (5), 3693–3700.

27. Nasser, F.; Lynch, I. Secreted Protein Eco-Corona Mediates Uptake and Impacts of Polystyrene Nanoparticles on Daphnia Magna. *J. Proteomics* **2016,** *137,* 45–51.

28. Wang, Z.; Wang, C.; Liu, S.; He, W.; Wang, L.; Gan, J.; Huang, Z.; Wang, Z.; Wei, H.; Zhang, J.; Dong, L. Specifically Formed Corona on Silica Nanoparticles Enhances Transforming Growth Factor β1 Activity in Triggering Lung Fibrosis. *ACS Nano* **2017,** *11* (2), 1659–1672.

29. Clift, M. J.; Bhattacharjee, S.; Brown, D. M.; Stone, V. The Effects of Serum on the Toxicity of Manufactured Nanoparticles. *Toxicol. Lett.* **2010,** *198* (3), 358–365.

30. Lesniak, A.; Campbell, A.; Monopoli, M. P.; Lynch, I.; Salvati, A.; Dawson, K. A. Serum Heat Inactivation Affects Protein Corona Composition and Nanoparticle Uptake. *Biomaterials* **2010,** *31* (36), 9511–9518.

31. Pattammattel, A.; Williams, C. L.; Pande, P.; Tsui, W. G.; Basu, A. K.; Kumar, C. V. Biological Relevance of Oxidative Debris Present in As-Prepared Graphene Oxide. *RSC Adv.* **2015,** *5* (73), 59364–59372.

32. Wong, C. H. A.; Sofer, Z.; Kubešová, M.; Kučera, J.; Matějková, S.; Pumera, M. Synthetic Routes Contaminate Graphene Materials with a Whole Spectrum of Unanticipated Metallic Elements. *Proc. Natl. Acad. Sci.* **2014,** *111* (38), 13774–13779.

33. Li, Y.; Boraschi, D. Endotoxin Contamination: A Key Element in the Interpretation of Nanosafety Studies. *Nanomedicine* **2016,** *11* (3), 269−287.

34. Ong, K. J.; MacCormack, T. J.; Clark, R. J.; Ede, J. D.; Ortega, V. A.; Felix, L. C.; Dang, M. K.; Ma, G.; Fenniri, H.; Veinot, J. G. Widespread Nanoparticle-Assay Interference: Implications for Nanotoxicity Testing. *PLoS One* **2014,** *9* (3), e90650.

35. Monopoli, M. P.; Walczyk, D.; Campbell, A.; Elia, G.; Lynch, I.; Baldelli Bombelli, F.; Dawson, K. A. Physical − Chemical Aspects of Protein Corona: Relevance to In Vitro and In Vivo Biological Impacts of Nanoparticles. *J. Am. Chem. Soc.* **2011,** *133* (8), 2525−2534.

36. Pattammattel, A.; Pande, P.; Kuttappan, D.; Basu, A.B.; Amalaradjou, A.; Kumar, C. V., Graphite exfoliation in animal sera: Scalable Production, Enhanced Stability and Reduced Toxicity. *Submitted Manuscript.*

37. Creighton, M. A.; Rangel-Mendez, J. R.; Huang, J.; Kane, A. B.; Hurt, R. H. Graphene-Induced Adsorptive and Optical Artifacts During In Vitro Toxicology Assays. *Small* **2013,** *9* (11), 1921−1927.

38. Hillegass, J. M.; Shukla, A.; Lathrop, S. A.; MacPherson, M. B.; Fukagawa, N. K.; Mossman, B. T. Assessing Nanotoxicity in Cells In Vitro. *Wiley Interdiscipl. Rev. Nanomed. Nanobiotechnol.* **2010,** *2* (3), 219−231.

39. Jarosz, A.; Skoda, M.; Dudek, I.; Szukiewicz, D. Oxidative Stress and Mitochondrial Activation as the Main Mechanisms Underlying Graphene Toxicity Against Human Cancer Cells. *Oxidat. Med. Cellular Longevity* **2015,** 2016, Article ID 5851035, 14 pages.

40. Delhaes, L.; Lazaro, J.-E.; Gay, F.; Thellier, M.; Danis, M. The Microculture Tetrazolium Assay (MTA): Another Colorimetric Method of Testing Plasmodium Falciparum Chemosensitivity. *Ann. Tropical Med. Parasitol.* **1999,** *93* (1), 31−40.

41. Strober, W. Trypan Blue Exclusion Test of Cell Viability. *Curr. Protoc. Immunol.* **2001,** A3. B. 1-A3. B. 3.

42. Borenfreund, E.; Puerner, J. A. Toxicity Determined In Vitro by Morphological Alterations and Neutral Red Absorption. *Toxicol. Lett.* **1985,** *24* (2-3), 119−124.

43. Gavrieli, Y.; Sherman, Y.; Ben-Sasson, S. A. Identification of Programmed Cell Death In Situ via Specific Labelling of Nuclear DNA Fragmentation. *J. Cell Biol.* **1992,** *119,* 493.

44. Gurtu, V.; Kain, S. R.; Zhang, G. Fluorometric and Colorimetric Detection of Caspase Activity Associated with Apoptosis. *Anal. Biochem.* **1997,** *251* (1), 98−102.

45. Frankfurt, O. S.; Krishan, A. Identification of Apoptotic Cells by Formamide-Induced DNA Denaturation in Condensed Chromatin. *J. Histochem. Cytochem.* **2001,** *49* (3), 369−378.

46. Wang, H.; Joseph, J. A. Quantifying Cellular Oxidative Stress by Dichlorofluorescein Assay Using Microplate Reader. *Free Radical Biol. Med.* **1999,** *27* (5), 612−616.

47. Akerboom, T. P.; Sies, H. Assay of Glutathione, Glutathione Disulfide, and Glutathione Mixed Disulfides in Biological Samples. *Methods Enzymol.* **1981,** *77,* 373−382.

48. Flohe, L. Superoxide Dismutase Assays. *Methods Enzymol.* **1984,** *105,* 93−104.

49. Hodges, D. M.; DeLong, J. M.; Forney, C. F.; Prange, R. K. Improving the Thiobarbituric Acid-Reactive-Substances Assay for Estimating Lipid Peroxidation in Plant Tissues Containing Anthocyanin and Other Interfering Compounds. *Planta* **1999,** *207* (4), 604−611.

50. Pop, C.; Salvesen, G. S.; Scott, F. L. Caspase Assays: Identifying Caspase Activity and Substrates In Vitro and In Vivo. *Methods Enzymol.* **2008,** *446,* 351–367.

51. Rafehi, H.; Orlowski, C.; Georgiadis, G. T.; Ververis, K.; El-Osta, A.; Karagiannis, T. C. Clonogenic Assay: Adherent Cells. *JoVE* **2011,** *49,* e2573-e2573.

52. Labarca, C.; Paigen, K. A Simple, Rapid, and Sensitive DNA Assay Procedure. *Anal. Biochem.* **1980,** *102* (2), 344–352.

53. Ahmed, S. A.; Gogal, R. M.; Walsh, J. E. A New Rapid and Simple Non-Radioactive Assay to Monitor and Determine the Proliferation of Lymphocytes: An Alternative to [3H] Thymidine Incorporation Assay. *J. Immunol. Methods.* **1994,** *170* (2), 211–224.

54. Regan, J. D.; Setlow, R.; Ley, R. Normal and Defective Repair of Damaged DNA in Human Cells: A Sensitive Assay Utilizing the Photolysis of Bromodeoxyuridine. *Proc. Natl. Acad. Sci.* **1971,** *68* (4), 708–712.

55. Harparkash, K.; HALLIWELL, B. Measurement of Oxidized and Methylated DNA Bases by HPLC with Electrochemical Detection. *Biochem. J.* **1996,** *318* (1), 21–23.

56. Schmid, W. The Micronucleus Test. *Mutat. Res./Environ. Mutagen. Related Sub.* **1975,** *31* (1), 9–15.

57. Clare, G. The In Vitro Mammalian Chromosome Aberration Test. *Genet. Toxicol. Principles Methods* **2012,** *817,* 69–91.

58. Vandghanooni, S.; Eskandani, M. Comet Assay: A Method to Evaluate Genotoxicity of Nano-Drug Delivery System. *BioImpacts: BI* **2011,** *1* (2), 87.

59. Doak, S. H.; Manshian, B.; Jenkins, G. J. S.; Singh, N. In Vitro Genotoxity Testing Strategy for Nanomaterials and the Adaptation of Current OECD Guidelines. *Mutat. Res./Genet. Toxicol. Environ. Mutagen.* **2012,** *745* (1–2), 104–111.

60. Zhang, Y.; Ali, S. F.; Dervishi, E.; Xu, Y.; Li, Z.; Casciano, D.; Biris, A. S. Cytotoxicity Effects of Graphene and Single-Wall Carbon Nanotubes in Neural Phaeochromocytoma-Derived PC12 Cells. *ACS Nano* **2010,** *4,* 3181–3186.

61. Zhou, H.; Zhang, B.; Zheng, J.; Yu, M.; Zhou, T.; Zhao, K.; Jia, Y.; Gao, X.; Chen, C.; Wei, T. The Inhibition of Migration and Invasion of Cancer Cells by Graphene via the Impairment of Mitochondrial Respiration. *Biomaterials* **2014,** *35* (5), 1597–1607.

62. Li, Y.; Liu, Y.; Fu, Y.; Wei, T.; Le Guyader, L.; Gao, G.; Liu, R.-S.; Chang, Y.-Z.; Chen, C. The Triggering of Apoptosis in Macrophages by Pristine Graphene Through the MAPK and TGF-Beta Signaling Pathways. *Biomaterials* **2012,** *33* (2), 402–411.

63. Sasidharan, A.; Panchakarla, L. S.; Sadanandan, A. R.; Ashokan, A.; Chandran, P.; Girish, C. M.; Menon, D.; Nair, S. V.; Rao, C. N. R.; Koyakutty, M. Hemocompatibility and Macrophage Response of Pristine and Functionalized Graphene. *Small* **2012,** *8* (8), 1251–1263.

64. Zhou, H.; Zhao, K.; Li, W.; Yang, N.; Liu, Y.; Chen, C.; Wei, T. The Interactions Between Pristine Graphene and Macrophages and the Production of Cytokines/Chemokines via TLR- and NF-κB-Related Signaling Pathways. *Biomaterials* **2012,** *33* (29), 6933–6942.

65. Schinwald, A.; Murphy, F. A.; Jones, A.; MacNee, W.; Donaldson, K. Graphene-Based Nanoplatelets: A New Risk to the Respiratory System as a Consequence of Their Unusual Aerodynamic Properties. *ACS Nano* **2012,** *6* (1), 736–746.

66. Teo, W. Z.; Sofer, Z.; Sembera, F.; Janousek, Z.; Pumera, M. Cytotoxicity of Fluorographene. *RSC Adv.* **2015,** *5* (129), 107158–107165.

67. Sawangphruk, M.; Srimuk, P.; Chiochan, P.; Sangsri, T.; Siwayaprahm, P. Synthesis and Antifungal Activity of Reduced Graphene Oxide Nanosheets. *Carbon. N. Y.* **2012,** *50* (14), 5156–5161.

68. Nguyen, H. N.; Castro-Wallace, S. L.; Rodrigues, D. F. Acute Toxicity of Graphene Nanoplatelets on Biological Wastewater Treatment Process. *Environ. Sci. Nano* **2017,** *4*, 160–169.

69. Yang, K.; Li, Y.; Tan, X.; Peng, R.; Liu, Z. Behavior and Toxicity of Graphene and Its Functionalized Derivatives in Biological Systems. *Small* **2013,** *9* (9–10), 1492–1503.

70. De Marzi, L.; Ottaviano, L.; Perrozzi, F.; Nardone, M.; Santucci, S.; De Lapuente, J.; Borras, M.; Treossi, E.; Palermo, V.; Poma, A. Flake Size-Dependent Cyto and Genotoxic Evaluation of Graphene Oxide on In Vitro A549, CaCo2 and Vero Cell Lines. *J. Biol. Regul. Homeost. Agents.* **2013,** *28* (2), 281–289.

71. Liu, Y.; Luo, Y.; Wu, J.; Wang, Y.; Yang, X.; Yang, R.; Wang, B.; Yang, J.; Zhang, N. Graphene Oxide Can Induce In Vitro and In Vivo Mutagenesis. *Sci. Rep.* **2013,** *3*, 3469.

72. Duch, M. C.; Budinger, G. R. S.; Liang, Y. T.; Soberanes, S.; Urich, D.; Chiarella, S. E.; Campochiaro, L. A.; Gonzalez, A.; Chandel, N. S.; Hersam, M. C., et al. Minimizing Oxidation and Stable Nanoscale Dispersion Improves the Biocompatibility of Graphene in the Lung. *Nano Lett.* **2011,** *11*, 5201–5207.

73. Lee, J. K.; Jeong, A. Y.; Bae, J.; Seok, J. H.; Yang, J.-Y.; Roh, H. S.; Jeong, J.; Han, Y.; Jeong, J.; Cho, W.-S. The Role of Surface Functionalization on the Pulmonary Inflammogenicity and Translocation into Mediastinal Lymph Nodes of Graphene Nanoplatelets in Rats. *Arch. Toxicol.* **2017,** *91* (2), 667–676.

74. Roberts, J. R.; Mercer, R. R.; Stefaniak, A. B.; Seehra, M. S.; Geddam, U. K.; Chaudhuri, I. S.; Kyrlidis, A.; Kodali, V. K.; Sager, T.; Kenyon, A., et al. Evaluation of Pulmonary and Systemic Toxicity Following Lung Exposure to Graphite Nanoplates: A Member of the Graphene-Based Nanomaterial Family. *Part. Fibre. Toxicol.* **2016,** *13*, 1–22.

75. Sawosz, E.; Jaworski, S.; Kutwin, M.; Hotowy, A.; Wierzbicki, M.; Grodzik, M.; Kurantowicz, N.; Strojny, B.; Lipińska, L.; Chwalibog, A. Toxicity of Pristine Graphene in Experiments in a Chicken Embryo Model. *Int. J. Nanomed.* **2014,** *9*, 3913.

76. Yang, K.; Zhang, S.; Zhang, G.; Sun, X.; Lee, S.-T.; Liu, Z. Graphene in Mice: Ultrahigh In Vivo Tumor Uptake and Efficient Photothermal Therapy. *Nano Lett.* **2010,** *10* (9), 3318–3323.

77. Singh, S. K.; Singh, M. K.; Nayak, M. K.; Kumari, S.; Shrivastava, S.; Grácio, J. J. A.; Dash, D. Thrombus Inducing Property of Atomically Thin Graphene Oxide Sheets. *ACS Nano* **2011,** *5*, 4987–4996.

78. Ahmadian, H.; Hashemi, E.; Akhavan, O.; Shamsara, M.; Hashemi, M.; Farmany, A.; Joupari, M. D. Apoptotic and Anti-Apoptotic Genes Transcripts Patterns of Graphene in Mice. *Mater. Sci. Eng. C* **2017,** *71*, 460–464.

79. Zhi, L.; Qu, M.; Ren, M.; Zhao, L.; Li, Y.; Wang, D. *Carbon. N. Y.* **2017,** *113*, 122–131.

80. Zanni, E.; De Bellis, G.; Bracciale, M. P.; Broggi, A.; Santarelli, M. L.; Sarto, M. S.; Palleschi, C.; Uccelletti, D. Graphite Nanoplatelets and Caenorhabditis elegans: Insights From an in Vivo Model. *Nano Lett.* **2012,** *12*, 2740–2744.

81. Dziewięcka, M.; Karpeta-Kaczmarek, J.; Augustyniak, M.; Rost-Roszkowska, M. *J. Hazard. Mater.* **2017,** *328*, 80–89.

Future of graphene revolution and roadmap

CHAPTER OUTLINE

10.1 Summary 207
10.2 Introduction 208
10.3 Graphene Manufacturing: A Critical Assessment 209
10.4 Health and Environmental Risk Assessment of Graphene 210
10.5 Graphene: Fundamental Education and Society 211
10.6 Conclusions 211
References 212

10.1 SUMMARY

A critical assessment of graphene science and suggestions to improve the benefit of graphene to society are discussed in this chapter. The lessons learned from carbon nanotube (CNT) research, for which the scientists are still evaluating the promises and pitfalls of the past, and our vision of graphene for a better future are also presented. First, we discuss the most important aspect of graphene technology: the large-scale manufacturing process. The need for a sustainable, scalable, and cost-effective method to produce highly pure graphene is anticipated by both the electronics and composite materials industries. Thorough characterization of the product to avoid or account for variabilities in the methods is necessary. Finally, the life cycle assessment of graphene that accounts for work place hazards, cytotoxicity, genotoxicity, and long-term impact on the environment is addressed. There is no question about the exciting properties or potential applications of graphene, but safety concerns about using this material in real life have to be accounted for prior to its widespread use, and to avoid past mistakes. In the concluding remarks, we discuss the socioeconomic impact of graphene in education, market investments, and job creation. Graphene has stimulated numerous new theories in fundamental science, which will have a place in textbooks in the near future. Careful investigations and standardized protocols in the field, as well

Introduction to Graphene. DOI: http://dx.doi.org/10.1016/B978-0-12-813182-4.00010-6

as responsible utility of graphene and its composite materials, are necessary for graphene in the service of society.

10.2 INTRODUCTION

Since the discovery of graphene in 2004, there has been an increased effort in academia and industry to fully harness its potential. The science of graphene has indisputably attracted the largest monetary investments for a single material in its time. About $150 million dollars in funding for graphene by the United Kingdom is an example of such initiatives.[1] The highest investors and patent holders of graphene technology are electronic companies such as Nokia and Samsung. In a 2015 report Nokia claimed that they have developed a proof-of-concept, transparent circuit using graphene.[2] One of the first graphene-based commercial products, released by a University of Manchester-based startup company, was found to be 10% more efficient than conventional LEDs.[3] Graphene was previously used to strengthen tennis rackets by a sporting goods company.[4] Such enterprises have opened up opportunities and competition across the world to manufacture superior and affordable graphene-based commodities in the market. In this chapter, we will discuss some of the challenges, practices, and promises in graphene science (Fig. 10.1) for future developments in this field.

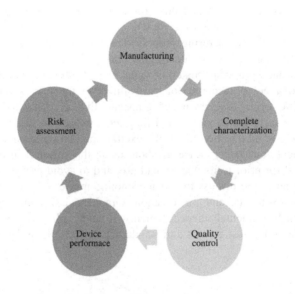

FIGURE 10.1

Proposed cycle of development in graphene science.

10.3 GRAPHENE MANUFACTURING: A CRITICAL ASSESSMENT

The major challenge in the advancement of graphene science is the development of synthetic methods that offer pure, defect-free, large-area graphene at an affordable price.[5] Chemical vapor deposition (CVD) methods have been cited as a way to address the problem of purity and pristine quality. However, these processes are time consuming and require skilled workers to transfer the graphene layer from metal substrates, a step that can affect mass production and expense.[6] At the time of writing, available graphene on various substrates produced by CVD methods costs about $5–$10 per mm^2 (sheet resistance 100–500 Ω^{-2}, price from different vendors). Graphene is considered a substitute for indium tin oxide (ITO) conductive thin films, which are priced at about $0.006–$1 per mm^2 (sheet resistance 15–20 Ω^{-2}, Sigma Aldrich price). This mismatch in price of graphene against ITO is a key problem to address. This implies that the need for inexpensive and sustainable synthetic methods for graphene is a current, unmet challenge.

Graphene made from graphite ($5–$10 per kg) is an alternative method to produce inexpensive graphene but the exfoliation process reduces the size (1–5 μm) and the product contains few layer graphene (1–5) but not necessarily single-layer graphene.[7] Commercially available powders of this type range from $100 to $500 per g, a price higher than that of competing carbon materials on the market (carbon black $1–$10 per g). This price range for graphene can be reduced in a commercial setting but there is a need for better methodologies. In the 1990s, CNTs were considered a magical material because of their unique features.[8] However, the excitement for CNTs eventually fizzled out due to their inability to meet expectations for a number of different reasons. Strikingly, the major reason for the diminished excitement was a lack of large-scale manufacturing methods to produce clean CNTs.[9] Sadly, researchers and funding agencies are moving away from CNT research, and experts now doubt that the commercialization of CNT products will ever become a reality. Therefore, it is essential to establish standard synthetic methods and quality assurance protocols in graphene research, quickly, before the optimism runs out. The recommendation in the field is to focus on nonoxidative routes to produce high-quality graphene that boosts the production rate as well as quality assurance, while its safety is being fully evaluated.

Oxidation of graphite to produce graphene oxide (GO) nano layers is the most explored route to produce graphene derivatives.[10] GO that can be easily synthesized in a standard laboratory setting using strong oxidants and then be dispersed in water attracted researchers in various fields, including biology. Despite the large number of publications on GO, many of its "amazing" qualities are found to be because of the impurities from synthesis steps.[11] These include electroactivity, photoluminescence, so-called "metal-free" catalysis, resistivity, and toxicity. These diversions suggest the need for nonoxidative routes for graphene production in solution, such as direct liquid-phase exfoliation of graphite. To date, graphite can be exfoliated with water, organic solvents, surfactants, proteins, and polymers (see Chapter 2: Synthetic routes to graphene preparation from the perspectives of possible biological applications), which are substitutable materials in applications that use GO or reduced GO. The advantage of these methods when compared to GO synthesis is the absence of oxidants (e.g., $KMnO_4/H_2SO_4$), which must be handled with maximum care during processing/purification and also at disposal, which would remarkably reduce the cost. For instance, the cost to produce few-layer, unoxidized graphene in our lab using graphite, with bovine serum albumin and a kitchen blender, was about $4−$6 per g (commercially available GO powders are $50−$100 per g).[12] Thus, the progress in graphene research depends critically on such cost-effective production of high-quality graphene with no toxic byproducts. However, the drawback of our method is that the samples contain the protein as the stabilizing agent which may not be desirable for some applications, while it may be excellent for other (biological) applications. Thus, a better and cleaner method that provides large-area graphene flakes in a nontoxic solvent that is of high quality and inexpensive is still urgently required.

10.4 HEALTH AND ENVIRONMENTAL RISK ASSESSMENT OF GRAPHENE

A major roadblock in responsible use of nanomaterials in day-to-day life, today, is a lack of good risk assessment studies because of large variations in size, heterogeneity, and other surface properties. These variabilities are not accounted for when different synthetic routes are adopted or different sources of graphene are assessed. The critics of CNT research argue that the fiber-like structure of CNT is as toxic as asbestos, although it has not yet been proven.[13] As discussed in Chapter 9, Nanotoxicity of graphene, toxicity studies on nanomaterials are often inconclusive and a test standard for nanomaterials is absolutely necessary. However, a few steps that can be taken to reduce the discrepancy in toxicity assays, from a production point of view, are discussed here. First, studies should focus on unoxidized samples, without any metallic reagent impurities. One of the drawbacks in the CNT toxicity study was the presence of catalytic impurities (Fe or Ni) from the growth process.[13] Acid treatments were used to remove this impurity from CNT but the oxidation of the carbon surface during this treatment changes the chemistry of the material. GO produced by the Hummers method may have trace amounts of manganese contamination, which could cause unwanted biological responses. Likewise, graphene produced by the CVD method can have contamination from a catalyst or the metal substrate.[14] Thus, cautious studies with metal-free graphene samples are necessary to assess the intrinsic toxicity of graphene.

A life cycle assessment of graphene is as important as acute/chronic toxicity tests. The fate of graphene inside the body and/or the environment will also have health, environmental, social, and economic impacts. Degradation of graphene by redox enzymes to produce small organic molecules is a major pathway but these species could also be toxic.[15] Attachment of graphene to environmental toxins or transition metal ions could be hazardous because of their dispersion in air or aquatic systems. Graphene is used in lower amounts in electronic devices than in composites, and therefore, toxicity and life cycle assessments should match with the direction of application as well. Reproducible results in graphene risk assessments are critical for further growth, which are often overlooked.

10.5 GRAPHENE: FUNDAMENTAL EDUCATION AND SOCIETY

The discovery of graphene has been a boost to both academic and industrial research. Isolation of freestanding graphene was a "textbook changing" breakthrough, which challenged the theoretical predictions that ruled out the existence of atomically thin layers. The unprecedented mechanical and electronic properties of graphene set many new boundaries in the physics and fundamental understanding of matter. Graphene has allowed for the observation of quantum hall effect, Dirac cone band structure, high carrier mobility because of their massless Fermion feature, etc.,[16] which allow for fruitful additions to fundamental physics and chemistry. The studies on graphene and its devices are still evolving and range from printed electronics to spintronics.[17]

Graphene is one of the most discussed materials by news media because of its potential applications in commonly used devices such as cell phones. This helps the growth of the graphene market and catalyzes public investment in graphene ($24 billion projected growth by 2025).[18] Increased media attention keeps the hope for this material at a higher level, but research studies require longer periods to establish the science on a firm footing. Some scientists claim that one reason for the fallout of CNT research is the high expectations in a short period of time. In this regard, scientists and society should have reasonable expectations about graphene to avoid a "CNT 2.0" type of situation from happening. All concerned parties can be optimistic but should also be cautious and realistic.

10.6 CONCLUSIONS

A critical analysis of graphene research progress is necessary for the development of graphene science that can benefit society. Careful and standardized protocols in manufacturing as well as life cycle assessment of graphene are required to fully harness its unique features (Fig. 10.2). The large investment and hopes from both government and private funding agencies are both catalysts for this field. The

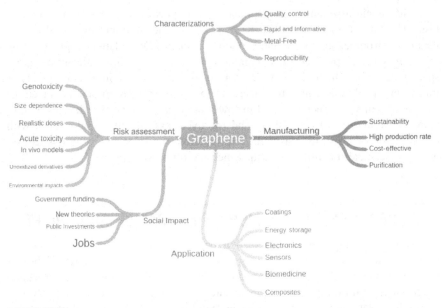

FIGURE 10.2

Current graphene science tree covers advancements and futuristic needs (created for free using coggle.it).

breakthroughs in this field continue to magnify the principles of physics, chemistry, and material sciences. New scientific knowledge stemming from graphene research at regular intervals keep revealing surprises from this "wonder material" in academia and industry. Collaborations across the field and industrial cooperation with academics could create more jobs, better products, and advanced science to benefit our society in the near future.

REFERENCES

1. https://www.theregister.co.uk/2017/02/14/graphen_busted_or_just_too_early/.
2. Colli, A.; Awan, S.A.; Lombardo, A.; Echtermeyer, T.J.; Kulmala, T.S.; Ferrari, A.C. Graphene-Based MIM Diode and Associated Methods. Google Patents: 2015.
3. http://www.manchester.ac.uk/discover/news/graphenes-lightbulb-moment/.
4. Lammer, H. Sporting Goods with Graphene Material. Google Patents; 2013.
5. Zurutuza, A.; Marinelli, C. Challenges and Opportunities in Graphene Commercialization. *Nat. Nano* **2014,** *9* (10), 730–734.
6. Ren, W.; Cheng, H.-M. The Global Growth of Graphene. *Nat. Nanotechnol.* **2014,** *9* (10), 726–730.
7. Nicolosi, V.; Chhowalla, M.; Kanatzidis, M. G.; Strano, M. S.; Coleman, J. N. Liquid Exfoliation of Layered Materials. *Science* **2013,** *340* (6139).

8. Pellegrino, E. M.; Cerruti, L.; Ghibaudi, E. M. Realizing the Promise—The Development of Research on Carbon Nanotubes. *Chem. Eur. J.* **2016,** *22* (13), 4330–4335.

9. http://cen.acs.org/articles/93/i23/Twists-Shouts-Nanotube-Story.html.

10. Hummers, W. S., Jr; Offeman, R. E. Preparation of Graphitic Oxide. *J. Am. Chem. Soc.* **1958,** *80* (6), 1339.

11. Wong, C. H. A.; Sofer, Z.; Kubešová, M.; Kučera, J.; Matějková, S.; Pumera, M. Synthetic Routes Contaminate Graphene Materials with a Whole Spectrum of Unanticipated Metallic Elements. *Proc. Natl. Acad. Sci.* **2014,** *111* (38), 13774–13779.

12. Pattammattel, A.; Kumar, C. V. Kitchen Chemistry 101: Multigram Production of High Quality Biographene in a Blender with Edible Proteins. *Adv. Funct. Mater.* **2015,** *25* (45), 7088–7098.

13. Kostarelos, K.; Bianco, A.; Prato, M. Promises, Facts and Challenges for Carbon Nanotubes in imaging and therapeutics. *Nat Nano* **2009,** *4* (10), 627–633.

14. Lupina, G.; Kitzmann, J.; Costina, I.; Lukosius, M.; Wenger, C.; Wolff, A.; Vaziri, S.; Östling, M.; Pasternak, I.; Krajewska, A., et al. Residual Metallic Contamination of Transferred Chemical Vapor Deposited Graphene. *ACS Nano* **2015,** *9* (5), 4776–4785.

15. Kotchey, G. P.; Hasan, S. A.; Kapralov, A. A.; Ha, S. H.; Kim, K.; Shvedova, A. A.; Kagan, V. E.; Star, A. A Natural Vanishing Act: The Enzyme-Catalyzed Degradation of Carbon Nanomaterials. *Acc. Chem. Res.* **2012,** *45* (10), 1770.

16. Geim, A. K.; Novoselov, K. S. The Rise of Graphene. *Nat. Mater.* **2007,** *6* (3), 183–191.

17. Han, W.; Kawakami, R. K.; Gmitra, M.; Fabian, J. Graphene Spintronics. *Nat. Nano* **2014,** *9* (10), 794–807.

18. Peplow, M. Graphene Booms in Factories But Lacks a Killer App. *Nature* **2015,** *522* (7556), 268.

Index

Note: Page numbers followed by "*f*" and "*t*" refer to figures and tables, respectively.

A

Absorption/extinction spectroscopy, 47–50
Antigen–antibody interactions, 156–157
Aqueous-phase exfoliation of graphite, 19
Arachidic acid, 31
Atomic force microscopy (AFM), 65–67, 77,
  81–82, 167

B

Ball milling of graphite, 24–25, 25*t*
Base washed GO–(bovine serum albumin,
  hemoglobin, or lysozyme), 161*t*
β-cyclodextrin/graphene, 148*t*
Bioconjugation with graphene derivatives,
  covalent approaches for, 160*t*
Biocorona formation, 189–190, 190*f*
Biofunctionalization of graphene derivatives by
  noncovalent chemistry, 161*t*
Biological activity studies, 171
Biological exfoliating agents, 32–38
Biological structures, conservation of, 172
Biomolecule–graphene interactions, 156–157
Bio–nano interface, 155–158, 189–190
Biophilized GO–catalase, 161*t*
Biophilized GO–cytochrome c, 161*t*
Biophilized GO–glucose oxidase, 161*t*
Biophilized GO–hemoglobin, 161*t*
Biophilized GO–horseradish peroxidase, 161*t*
Biosensing, 177–179
  electrochemical graphene sensors, 178–179
  fluorescence-based graphene biosensors,
    177–178
Block copolymer/graphene, 145*t*, 148*t*
Boron nitride (BN), 8–9, 82
Borophene, 91–92
Bottom-up approaches, 19–23
  chemical synthesis method, 21–23
  chemical vapor deposition (CVD) method,
    20–21, 20*f*
Bound enzymes, preservation of structure and
  function of, 172–174
Bromodeoxyuridine (BrdU) assay, 193*t*

C

Caenorhabditis elegans, 192
Carbon allotropes, graphene hybrids with, 123
  graphene/CNT hybrids, 125–129

covalent CNT/graphene hybrids, 125–126
noncovalent CNT/graphene hybrids, 127–129
graphene/fullerene hybrids, 129
organic chemistry of graphene, 129–134
Carbon nanotubes (CNTs), 5–6, 123–124, 209,
  211
  schematic structures of, 6*f*
Caspase-1 assay, 193*t*
Caspase-3 activity, 193*t*
Cationized BSA (cBSA), 164, 167, 168*f*, 172–174
Cetyltrimethyl ammonium bromide, 31
Characterization techniques for graphene, 45
  microscopy methods, 63–67
    atomic force microscopy/scanning tunneling
      microscopy, 65–67
    fluorescence quenching microscopy (FQM),
      67
    scanning electron microscopy (SEM) and
      transmission electron microscopy (TEM),
      63–65
  spectroscopic methods, 46–63
    absorption/extinction spectroscopy, 47–50
    IR spectroscopy, 55–56
    Raman spectroscopy, 50–55
    surface plasmon resonance (SPR)
      spectroscopy, 61–63
    X-ray diffraction (XRD), 57–59
    X-ray photoelectron spectroscopy, 56–57
    zeta potential measurements, 59–61
Chemical synthesis method, 21–23
Chemical vapor deposition (CVD), 8–9, 17–18,
  20–21, 20*f*, 46, 76, 103–105, 125, 209
Chemically modified BSA, 172
Chlorosulfuric acid, 29
Chromosomal aberration test, 193*t*
Circular dichroism (CD) spectroscopy, 163,
  168–170, 170*f*
Clonogenic assay or colony-forming efficiency
  (CFE), 193*t*
Comet assay (single-cell gel electrophoresis,
  SCGE), 193*t*
Conjugated polymer-graphene hybrids, 149
Covalent chemistry, 125, 147
  bioconjugation via, 160*t*
Covalent CNT/graphene hybrids, 125–126
Critical assessments, 209–210
Crosslinked silicon polymer/graphene, 148*t*
Cubic form of BN (c-BN), 8–9

CVD graphene—bovine serum albumin, 160*t*
N-Cyclohexyl-2-pyrrolidone, 92—93
Cytochrome c (Cyt c), 164—165, 172—173

D

DCF (dichlorofluorescein) assay, 193*t*
Defect-free graphene, 17—18, 21, 26—27
Degradation of graphene, 211
Development in graphene science, 208*f*
Dichalcogenides, 1, 83—86
    applications, 86
    characterization techniques, 85—86
    exfoliation, 83—85
Diels—Alder reactions, 131—132
N, N'-Dimethyl-2,9-diazaperopyrenium dication,
        30
Dimethyl formamide (DMF), 29
2-(Dimethylamino) ethyl methacrylate (MAEMA),
        29
N-[3-(Dimethylamino)propyl]methacrylamide
        (DMAPMA), 29
Discovery of graphene, 1
    boron nitride (BN), 8—9
    comparisons with other carbon-based
        nanomaterials, 5—8
        exfoliated graphene, 8
        graphene oxide (GO), 6
        reduced graphene oxide (rGO), 6—7
    future advances and challenges, 11—12
    history, 2—3
    MXenes, 10—11
    phosphorene, 11
    terminology, 4—5
    transition metal dichalcogenides (TMDs), 9—10
    wonder material, 4
DNA content assay, 193*t*
DNA laddering, 193*t*
DNA/GO, 148*t*

E

Electrochemical exfoliation of graphite, 25—26,
        26*t*
Electrochemical graphene sensors, 178—179
Electron energy-loss spectroscopy (EELS), 109
Electron microscopy, 63, 77
Energy-dispersive electron spectroscopy (EDX),
        109
Enzymatic activities, measuring, 171, 171*f*
Enzymes
    kinetic stability of, 175—176
    thermodynamic stabilities of, 174—175
Epitaxial graphene, 3
Epoxy resin/graphite, 145*t*

Epoxy/graphene, 148*t*
Equilibrium binding studies, 163—165
Ethyl cellulose/graphene, 145*t*, 148*t*
Exfoliated graphene, 8
Exfoliating agents, 27—38
    biological, 32—38
    nonbiological, 29—32
        organic solvents as exfoliating agents, 29—30
        surfactants as exfoliating agents, 30—32
Exfoliation methods, 19, 46
    comparison of, 27
Exfoliation of graphite, 27*t*, 30, 33, 144
    biomolecules used for, 34*t*
    shear- and turbulence-based, 27*f*

F

Fluorescein isothiocyanate (FITC), 165
Fluorescence quenching microscopy (FQM), 67
Fluorescence spectroscopy, 46—47, 165—166
Fluorescence-based graphene biosensors, 177—178
Fourier transform infrared (FTIR), 46—47
Fullerenes, 5—6, 123—124, 129
    schematic structures of, 6*f*
Fundamental science, 129, 207—208, 211
Future of graphene revolution and roadmap, 207,
        212*f*
    fundamental education and society, 211
    graphene manufacturing, 209—210
    health and environmental risk assessment of
        graphene, 210—211

G

Genotoxicity, 192
Global market for graphene, 4
Glucose oxidase (GOx), 164—165
Glucose-detecting devices, 178—179
Glutathione/oxidized glutathione (GSH/GSSG)
        assay, 193*t*
Graphene biohybrids
    biocatalysis using graphene platform, 171—176
        kinetic stability of enzymes, 175—176
        preservation of structure and function of
            bound enzymes, 172—174
        thermodynamic stabilities of enzymes,
            174—175
    biosensing, 177—179
        electrochemical graphene sensors, 178—179
        fluorescence-based graphene biosensors,
            177—178
    characterization of composites, 162—171
        atomic force microscopy, 167
        biological activity studies, 171
        CD spectroscopy, 168—170

equilibrium binding studies, 163–165
  fluorescence spectroscopy, 165–166
  X-ray diffraction, 168
  zeta potential studies, 166–167
design and preparation of, 159–161
  covalent and noncovalent approaches, 159–161
drug delivery applications, 179
Graphene oxide (GO), 4–6, 103–104, 125, 142–143, 165–166, 209–210
  formation of, from graphite, 7f
  GO–(GOx1HRP)-polyacrylic acid, 161t
  GO–Angiopep-2, 160t
  GO-based hybrids, 147
  GO-based models, 188–189
  GO–bilirubin oxidase, 161t
  GO–bovine serum albumin (BSA), 160t
  GO–BSA, 161t
  GO–cationized glucose oxidase, 161t
  GO–chitosan, 160t
  GO–chymotrypsin, 161t
  GO–Cyanobacterium metallothionein (SmtA), 160t
  GO–DA-grafted heparin, 160t
  GO–fibrin, 161t
  GO–gelatin, 161t
  GO–glucose oxidase (GO/GOx), 160t, 161t
    stability of, 175–176
  GO–hemoglobin, 161t
  GO–horseradish peroxidase, 161t
  GO–laccase, 160t
  GO–lipase, 160t
  GO–lysozyme, 161t
  GO–PEG/trypsin, 161t
  GO–tyrosinase, 161t
  hybrid systems, 155–156
  representative structure of, 5f
  synthesis via Hummer's method, 190
  toxicity, 187–188
Graphene revolution, 4
Graphene/polymer composites, 141
  biomedical application of, 150
  in situ synthesis of, 144–147
    characterization, 144–146
    design and preparation, 144
    properties and applications, 145t, 146–147
  modification routes for, 143f
  postmodification approaches, 147–149
    characterization, 147–149
    design and preparation, 147
    properties and applications, 148t, 149
Graphite, 5–6, 8, 209
  biomolecules used for exfoliation of, 34t
  electrochemical exfoliation of, 25–26, 26t
Graphite exfoliations, 23–24, 30, 30f, 32, 39t

H
[³H] thymidine assay, 193t
Health and environmental risk assessment of graphene, 210–211
Hexagonal boron nitride (h-BN), 1, 8–9, 9f, 77–82, 103–104
  applications of, 82
  aqueous exfoliation of, 80t
  characterization techniques, 79–82
  graphene composites with, 105–110
    characterization, 108–109
    design and preparations, 105–108
    properties and applications, 110
  production, 78–79
Highly oriented pyrolytic graphite (HOPG), 65
High-resolution transmission electron microscopy (HR TEM), 63, 81–82
History, of graphene, 2–3
Hummers method, 6, 7f
Hybrid materials, 134–135, 141
  synthesis and application of, 130t

I
3,3′-Iminobis (N,N-dimethylpropylamine) (DMPA), 29
In vitro toxicity of graphene, 192–197, 196t
In vivo toxicity of graphene, 198, 199t
Inductively coupled plasma–mass spectrometry/optical emission spectroscopy (ICP-MS/ICPOES), 191
Infrared (IR) spectroscopy, 55–56, 163
Inorganic 2-D materials, graphene composites with, 103
  hexagonal boron nitride (h-BN), graphene composites with, 105–110
    characterization, 108–109
    design and preparations, 105–108
    properties and applications, 110
  molybdenum disulfide (MoS₂), graphene composites with, 110–114
    characterization, 112
    design and preparation, 110–112
    properties and applications, 113–114
  nongraphene heterostructures, 117
  transition metal dichalcogenides (TMDs), composites with, 114–116
Inorganic analogues of graphene, 75
  hexagonal boron nitride, 77–82
    applications of h-BN, 82
    characterization techniques, 79–82
    production, 78–79
  monoatomic layers, 91–94
    applications, 94

Inorganic analogues of graphene (*Continued*)
  characterization, 93
  synthesis or exfoliation, 92–93
  MXenes, 87–91
    applications, 91
    characterization techniques, 90
    production, 88–90
  transition metal dichalcogenides, 83–86
    applications, 86
    characterization techniques, 85–86
    exfoliation, 83–85
In-plane heterostructures, 103–105
Investments for graphene, 208
Ion-coupled binding (ICPB) mechanism,
    157–158

L

Lactate dehydrogenase (LDH) assay, 193*t*
Langmuir–Freundlich isotherm, 163, 164*f*
LASER scribbling, 6–7
Liquid-phase exfoliation of graphene, 11, 17–18,
    23–27, 46–47, 76–77
  ball milling, 24–25
  common exfoliation methods, comparison of, 27
  electrochemical exfoliation, 25–26
  shear exfoliation, 26–27
  ultrasonication, 24

M

Malondialdehyde (MDA) assay, 193*t*
Mass production, 49–50, 129, 147, 209
Measurement of oxidized guanine bases, 193*t*
Mechanical peeling, 23
Microarrays, 193*t*
Microculture tetrazolium assays (MTA), 193*t*
Micronucleus test, 193*t*
Microscopy methods, 63–67
  atomic force microscopy/scanning tunneling
      microscopy, 65–67
  fluorescence quenching microscopy (FQM), 67
  scanning electron microscopy (SEM) and
      transmission electron microscopy (TEM),
      63–65
Molecular beam epitaxy (MBE) process, 105
Molecular graphene, 19–20
Molybdenum disulfide (MoS_2), graphene
    composites with, 110–114
  characterization, 112
  design and preparation, 110–112
  properties and applications, 113–114
Molybdenum sulfide, 1, 10*f*
  applications of MoS_2 nanosheets, 87*t*
  aqueous exfoliation of, 84*t*

Monoatomic layers, 91–94
  applications, 94
  characterization, 93
  synthesis or exfoliation, 92–93
MXenes, 10–11, 87–91
  applications, 91
  characterization techniques, 90
  production, 88–90

N

Nanocellulose, 31
Nano-environmental health and safety (nano-EHS),
    188
Nanosafety, 187–188, 198–201
Nanotoxicity of graphene, 187
  challenges in experimental design of
      nanotoxicity assays, 189–191
  in vitro toxicity of graphene, 192–197, 196*t*
  in vivo toxicity of graphene, 198, 199*t*
  methods for assessments, 192, 193*t*
Neutral red assay, 193*t*
N-methyl pyrrolidine (NMP), 8, 27–29, 59, 83
Nokia, 208
Nonaromatic surfactants with long aliphatic
    chains, 31
Nonbiological exfoliating agents, 29–32
  organic solvents as exfoliating agents, 29–30
  surfactants as exfoliating agents, 30–32
Noncovalent chemistry, 123–124
  biofunctionalization of graphene derivatives by,
      161*t*
Noncovalent CNT/graphene hybrids, 127–129
Noncovalent hybrids, 147
Nongraphene heterostructures, 103–104, 117
Nonionic aromatic surfactants, 30–31
Nontargeted interactions at bio–nanosheet
      interface, 158*f*
Northern blot analysis, 193*t*
Nuclear magnetic resonance (NMR) techniques,
    124–125

O

Oligomeric ionic liquid/graphene, 145*t*
Organic chemistry of graphene, 129–134
Organic reactions of pristine graphene, 131*f*, 132*t*
Organic solvents as exfoliating agents, 29–30
Ortho-dicholorobenzene (*O*-DCB), 29
Oxidative debris, 190

P

PCR (polymerase chain reaction) arrays, 148*t*, 193*t*
Perylenebisimide bolaamphiphile, 30–31
Phosphorene, 11, 93

π−π interaction, 129
Poly(3-hexylthiophene)/graphene, 145t
Poly(acrylic acid), 148t, 174−175
Poly(vinylpyrrolidone)/graphene, 145t
Poly[(2-(methacryloyloxy)ethyl)
    trimethylammonium chloride] (PMETAC)/
    graphene, 148t
Polycarbonate/graphene, 148t
Polyethylene glycol (PEG)-modified graphene
    derivatives, 142−143, 148t
Polyethylene/graphite, 145t
Polymer-enhanced graphite exfoliation, 144
Polymeric surfactants with aromatic side groups
    and hydrophilic groups, 32
Polymethylmethacrylate (PMMA), 21
Polystyrene/graphene, 145t, 148t
Polystyrene/graphite, 145t
Postmodification of graphene, 147−149
Postsynthetic synthesis, 142−143
Powder XRD, 59, 90, 168, 169f
Pristine graphene, 4−5, 60, 129−130, 131f, 143
    organic reactions of, 131f
Protein-exfoliated graphene, 36−37
    storage stability of, 37f
Pyrene functionalized block copolymer/graphene,
    145t
Pyrene-DNA/graphene, 145t
Pyrene-modified monomers, 144
Pyrene-modified polymers, 149f
Pyrene-polycaprolactone, 32
Pyrene-polytheleneglycol, 32
1,3,6,8-Pyrenetetrasulfonic acid, 30

Q

Quantitative real-time polymerase chain reaction
    (qRT-PCR), 193t
Quinquethiophene-terminated PEG/graphene, 145t

R

Raman spectroscopy, 36−37, 46−47, 50−55,
    57−58, 63−64, 79−81, 93, 93f, 129,
    132−134
Reactive nitrogen species (RNS), 188
Reduced graphene oxide (rGO), 4−7
    -based hybrids, 141−142
    representative structure of, 5f
    rGO−Aptamer, 161t
    rGO−DNA, 161t
    rGO−Horseradish peroxidase, 161t
    rGO−lipase, 161t
    rGO−oxalate oxidase (OxOx), 161t
Representative structure of graphene, 5f
Ribonuclease protection assays (RPA), 193t

S

Samsung, 208
Scanning electron microscopy (SEM), 63−65,
    125−126
Scanning tunneling electron microscopy (STEM),
    109
Scanning tunneling microscopy, 3f, 65−67
Scotch tape method, 2−3, 19, 23, 92−93
Shear exfoliation, 26−27
Single-layered graphene (SLG), 1, 3−4, 61−62
Small-angle X-ray spectroscopy (SAXS), 168
Socioeconomic impacts, 207−208
Sodium cholate, 31
Sodium dodecylbenzene sulfonate (SDBS), 30
Spectroscopic methods, 46−63
    absorption/extinction spectroscopy, 47−50
    IR spectroscopy, 55−56
    Raman spectroscopy, 50−55
    surface plasmon resonance (SPR) spectroscopy,
        61−63
    X-ray diffraction (XRD), 57−59
    X-ray photoelectron spectroscopy, 56−57
    zeta potential measurements, 59−61
"Stable-on-the-table" enzyme hybrids, 155−156
Standardized protocols, 207−208, 211−212
Staudenmeir−Hoffman−Hamdi oxidation method,
    6
Sterilization of the nanoparticles, 191
Stern−Volmer analysis, 165−166
Stimuli-responsive drug-delivery systems, 179
Storage stability of protein-exfoliated graphene, 37f
Streptavidin-avidin, 156−157
Sugar-lectin, 156−157
Superoxide dismutase (SOD) assay, 193t
Surface plasmon resonance (SPR) spectroscopy,
    61−63
    SPR angle, 61
Surfactants as exfoliating agents, 30−32
Synthetic routes to graphene preparation, 17
    bottom-up approaches, 19−23
        chemical synthesis method, 21−23
        chemical vapor deposition (CVD) method,
            20−21, 20f
    top-down approaches, 23−38
        exfoliating agents, 27−38
        liquid-phase exfoliation of graphene, 23−27
        mechanical peeling, 23

T

TdT dUTP nick end labeling (TUNEL) and
    apostain assays, 193t
2-(Tert-butylamino) ethyl methacrylate (BAEMA),
    29

Thermogravimetric analysis (TGA), 147—149
Titanium dioxide, 188
Top-down approaches, 23—38
    exfoliating agents, 27—38
        biological exfoliating agents, 32—38
        nonbiological exfoliating agents, 29—32
    liquid-phase exfoliation of graphene, 23—27
        ball milling, 24—25
        comparison of common exfoliation methods, 27
        electrochemical exfoliation, 25—26
        shear exfoliation, 26—27
        ultrasonication, 24
    mechanical peeling, 23
Transition metal dichalcogenides (TMDs), 8—10, 75—76, 83—86, 103—104
    applications, 86
    characterization techniques, 85—86
    composites with, 114—116
    exfoliation, 83—85
Transmission electron microscopy (TEM), 3, 3f, 36—37, 63—65, 109, 129
TritonX-100, 31
Trypan blue exclusion assay, 193t

U
Ultrasonication, 8, 17—19, 24

V
Vinylimidazole-based polymer/graphene, 145t

W
White graphene. See Hexagonal boron nitride (h-BN)
Wonder material, graphene as, 1, 4

X
X-ray diffraction (XRD), 57—59, 168
X-ray photoelectron spectroscopy (XPS), 46—47, 56—57

Z
Zeta potential, 52f, 59—61
    measurements, 59—61
    studies, 166—167
Zinc oxide, 188